Religion and the Sciences

Man and Religion Series
Part 7/
Christianity and Our Culture

The Religious Education Press Ltd
a member of the Pergamon Group
Headington Hill Hall
Oxford

Religion and the Sciences

Keith Wilkes

Keith Wilkes
is
Chaplain and Lecturer in Theology
College of St Matthias
Bristol

Design and Cover
Keith Clements

Man and Religion Series
General Editor Ronald Dingwall
Executive Editor Gordon Hawes MA BD
Art Director Keith Clements

The Religious Education Press Ltd
a member of the Pergamon Group
Oxford London Edinburgh
New York Toronto Sydney

First published 1969

Made and printed in Great Britain by
A Wheaton and Company
Exeter Devon

Library of Congress Catalog Card No. 73-82387

08 006567 8 (flexicover)
08 015593 6 (hard cover)

Contents

Introduction

In the study of contemporary culture few aspects are more fascinating than the development of modern science, and few things are more important for responsible people to understand, since the scientific attitude is shared—often unknowingly—by more and more people, and its influence is shaping human thought and life to an ever increasing extent. For these reasons alone, a study of the origins and nature of the scientific revolution deserve serious attention in schools and colleges. But for religious people there is an added incentive, since modern science has emerged from a civilization permeated by the Christian faith, and to many appears to threaten the roots of belief: it certainly demands a great deal of rethinking about the nature of religion, and its relationship to scientific knowledge.

In spite of the nature of the subject, there is evidence of students' reluctance to undertake scientific studies at school or university: a reluctance encouraged by the bad image presented by the use (or misuse) to which scientific knowledge is put. The shortage of science specialists in schools, and a certain unease on the part of school (and some college) staffs to accord the sciences—and technology—an equal place in academic estimation with the humanities, have all contributed to a shortage of young people pursuing scientific studies. The assumption that scientists cannot be Christians is widely held, and the problems raised by modern science for religious faith are often inadequately met in sixth form or college discussion.

This book has been prepared with this situation in mind. It is designed for several possible uses: as a student's reader; as a class text providing an elementary introduction to modern discussions of the history and philosophy of science in relation to religion; as a source book for R.E. and General Studies teachers. It is also intended to serve as a handbook for basic theology and combined subject courses in Colleges of Education, and for liberal studies courses in Universities and Colleges of Technology.

In the approach adopted here, an attempt has been made to set the debate between science and religion in its historical and cultural context. At the same time, some attention has been given to the actual processes of experiment and thought in science alongside reference to some of the key personalities of the scientific revolution. It seems to me necessary to give the reader an idea of the 'feel' of science in this way, as well as a sketch of its conceptual conclusions, if the full impact of modern science is to be appreciated. It is not an unreasonable assumption that many of those concerned with courses in religious education and liberal studies are not very familiar with the history of science presented in this way.

Certain parts of the book are frankly philosophical in an elementary way. This is quite deliberate, since these discussions are usually the most enjoyable for the student and of the essence of the debate between the sciences and religion. It is my conviction that the sixth-former as well as the college student needs to come to grips with philosophical issues—especially if the normal curriculum makes no provision for courses in philosophy, as is unhappily very often the case. The questions asked by students in school and college are far more frequently of a philosophical character than is sometimes realized, and ought to be dealt with as such. It is hoped that the brief outline of certain key themes in modern philosophy will encourage the student and teacher to extend their enquiries further.

The suggestions for questions, projects and discussion at the end of each chapter are by no means exhaustive. They can be used in a variety of ways. Some have been included for fun. They are made as a result of some experience in dealing with this field of enquiry in the past ten years among sixth formers and students in different institutions of higher education. They form an integral part of the book, which is designed to raise questions and open up new fields of discovery rather than provide final answers. Used for this purpose, I hope this book—with its companion volume **Religion and Technology**—will serve as a modest guide in the search for truth and understanding for which Christianity, as I believe, provides the central clues.

Acknowledgements

The author wishes to acknowledge his indebtedness to the various authors and publishers noted in the text and lists of sources. In particular, he is indebted to two major studies published in recent years covering the field surveyed briefly in this volume: **The Scientific Revolution** by A. R. Hall (Longmans), and **Issues in Science and Religion** by I. G. Barbour (S.C.M. Press; U.S.A. © 1966 Prentice Hall, Inc., reproduced by permission. Acknowledgement is also made to Professor R.A. Lyttleton for the quotation on pp. 97–98 from his book **The Modern Universe**, (Hodder and Stoughton; U.S.A. Harper and Row).

In addition, the author wishes to record his gratitude to several friends and colleagues for their helpful criticism and encouragment; especially to Ronald Dingwall (General Editor of the "Man and Religion" Series), the Reverend W. David Stacey (Wesley College, Bristol), the Reverend Gordon Pavey and Mr. David Burch (College of St. Matthias, Bristol), Mrs. Sybil Hodge and Dr. John Tesh (University of Bristol) and several groups of students at the University of Birmingham and the College of St. Matthias, Bristol, on whom ideas contained in this book have been inflicted in teaching and discussion. He has frequently accepted their corrections and followed their advice, but none of them is to be held responsible for the imperfections that remain.

He would like to thank Mrs. J. L. Atkinson who typed the manuscript with speed and accuracy. Finally, he wishes to express his continuing gratitude to his wife, to whom this modest volume is dedicated, and without whose support and encouragement in countless ways, this undertaking would never have been completed.

Part One
THE RISE OF MODERN SCIENCE

A

From the Medieval Cosmos to the Newtonian World-Machine

1

The Background of the Middle Ages

It may seem a far cry to link the name of Geoffrey Chaucer (c. 1340–1400) with the origins of modern science, but in point of fact that remarkable poet and observer of the late fourteenth century scene can stand as a kind of prologue to the fascinating story of the rise of the scientific world. His acute comments in the **Canterbury Tales** indicate many important elements in the background from which this new movement emerged in western Europe. He notes the extent of trade and travel, especially to Italy which was the original centre of intellectual activity from the twelfth to the fifteenth century. He shows something of the northward shift of that activity to Paris as the peoples of northern Europe became more prosperous and better fed. He suggests that the rise of the Turkish Empire was a matter of great consequence—as indeed it was, severing the West's trade with and cultural dependence upon the East. He points to the influence of Aristotle's thought in scholarship (the Clerk from Oxford), with which Christendom had to come to terms or lose its integrity.

Chaucer's world, as anyone reading Neville Coghill's translation of **The Canterbury Tales** (1951) will know, is a vital, full-blooded, colourful affair. In spite of wars and plagues, it appears as a stable, secure stage peopled with tough, earthy characters. It conveys the impression of bustling variety, human involvement and interest, hard work and humour. Whether we look at the Miller, the Knight, the Monk or that fascinating character the woman from Bath, the impression of an age of colour, vitality, and humanity is the same: you can almost smell what it was like! But what gave this wholesome consistency to the Middle Ages was the penetration and integration of the whole scene by religion. The idea of the pilgrimage in **The Canterbury Tales**, is more than a literary device, it is an allegory of the human situation. The purpose and providence of God touches every concern, and upholds the marvellous variety of life and character. It sustains the social order with knight and peasant, merchant and monk, priest and layman, man and woman. In this wholesome, realistic, bawdy world, religion is accepted as the essential ingredient by everyone, and the pilgrimage is an allegory of everyone's response. Piety is the most natural thing; the journey to Canterbury is inevitable; the tales are brilliant cameos of character, wit and human understanding; Chaucer's method and style are the climax of a culture embracing God, men and nature in a splendid whole.

Yet Chaucer is the sign of a new era. The detachment which enables him to observe and describe so vividly the Medieval scene places him outside it. Many of the stories he retells come from outside the Medieval arena. His literary style and awareness of human nature place him as a man of the Renaissance.

Chaucer was not only a distinguished man of letters and administration, he was also something of an expert in contemporary sciences: astronomy, medicine, physics, and alchemy. He drew on his knowledge in several of his stories. The **Prologue** to the **Canterbury Tales** mentions Islamic medical authorities of the ninth century still accepted in the fourteenth; the **Canon's Yeoman's Tale** shows a detailed and accurate knowledge of alchemy; his allegory **The House of Fame,** written *circa* 1375, contains a long account of sound waves. In his translation of an Arabic text about the astrolabe—an instrument used by Greek astronomers for the measurement of the angles of stars in relation to a fixed line—Chaucer took part himself in the process of making available to his contemporaries the intellectual and practical riches of the ancient world. It was this rediscovery of the heritage of Greece, supplemented by the additions made by Arabic scholars, which was to prove of the greatest importance to the scientists of the Middle Ages, as well as to the later humanists of the Renaissance of the fifteenth and sixteenth centuries. In particular the thought of the Hellenists was wider in range and more fearless in logical rigour than that passed on through the so-called Dark Ages. Its influence was felt first in the springs of the 'little Renaissance' of the twelfth century which rested on the fruits of Arabic scholarship; it was to become a full flood by the end of the fifteenth.

The inflow of ideas from China and the East had been considerable in the thirteenth century. It brought the halter and the heavy iron plough which did so much to raise the level of food production, health and prosperity in the farmlands of Britain and Northern Europe. It introduced rag paper and block printing, and possibly the use of the windmill. The rise of the Turks effectively blocked the trade routes, pioneered by men like Marco Polo (1250–1323), through which many of these cultural innovations were received. Thereafter the West was obliged to rediscover its heritage for itself and rely on its own inventive genius. When the routes to the East were reopened, the cultural traffic passed the other way.

In the twelfth century the West had access to the works of Arabic scholars, many of which were versions of classical writings. Large numbers of these works were translated by men like Gerard of Cremona (c. 1114–87) who made available in Latin as many as ninety-two complete treatises including Ptolemy's **Almagest**, Euclid's **Elements**, the **Canons**—a medical encyclopaedia—of Avicenna and many others. Another of these energetic translators of Arabic scholarship was an Englishman, Adelard of Bath, who had disguised himself as a Mohammadan student and attended lectures at Cordoba, then made available his knowledge of Arabic science in a book called **Natural Questions** (c. 1100) and translated the **Arithmetic** of al-Kirismi containing the Indian system of number.

From the fifteenth century onwards, European scholars went directly to the ancient texts. Huge lexicons were compiled and the task of translation from Greek and Hebrew as well as Arabic was considerably accelerated. This was aided by the Turks' capture of Byzantium when many of the treasures of its library were scattered throughout Europe, and by the development of the printing press. Although religious and literary works were the first to be printed (the Bible in 1454), soon it was the turn of ancient scientific writing such as Pliny's **Natural History** (1469), Ptolemy's **Geographica** (1475), the three biological works of Aristotle (1476), Euclid (1482) and later works by Ptolemy and Archimedes (1528–44).

In spite of some ecclesiastical opposition to such pagan studies, it was not long before the 'new' learning spread and was widely taught at the recently founded universities.

Men like Robert Grosseteste at Oxford (1175–1253) soon realized the importance of these classical writings. He himself became a leading exponent of Aristotle, emphasizing the importance of observation and experiment in scientific understanding.

Strangely enough, this aspect of Aristotle's thought was completely neglected by Thomas Aquinas (1224–74) in the second half of the thirteenth century. For this reason, official Christendom—which adopted Aquinas's great synthesis—was to remain particularly cautious about, and frequently hostile to, the growth of experimental science. Aquinas's work of uniting the substance of Christian faith with the categories and framework of Aristotle's philosophy is chiefly contained in his two major works: **Summa Contra Gentiles** written between 1259 and 1264, and **Summa Theologica** written from 1325 and published after his death.

Within a remarkably short space of time, the scholars of the

Christian West, like the late Medieval engineers, had assimilated the 'new' ideas of the ancient world, and reached a point at which their own critical enquiry and exploration could begin. By the beginning of the fourteenth century there are already signs of true science in the West, greatly assisted by the use of mathematics derived from the Arabs—though originating long ago in Babylon and India. Anatomy, mechanics, geometry developed apace in places like Padua; new technical ideas such as the magnetic compass, distillation and the mechanical clock appeared; and a new attitude towards commercial efficiency and the natural world began to take shape. All this occurred as printing with block and moveable type began—and long before the full flowering of Renaissance humanism in the sixteenth century.

The intellectual life of late Medieval Europe which rested on the twelfth century 'little renaissance' had a surprising unity, but it would be a mistake to suppose that it was uniform. Aristotle (and Aquinas) were not without their critics and opponents, such as Nicole Oresme in Paris, the Averroists who swept the University of Paris with their scepticism in the thirteenth century, and the disciples of William of Ockham at Oxford after them. Nevertheless, such was the power of Aristotle's logic and method, and so attractive were his concepts to the Medieval mind, that the terms of the scholastic debates and the problems considered were those which Aristotle's scheme suggested. The work of these critics usually appeared in their comments on Aristotle's texts rather than from original propositions or books of their own. Thus it is necessary to consider the main tenets of the Aristotelian scheme—with St. Thomas's additions—to grasp the world-view at the start of the scientific era.

2
The Medieval Scheme

The Greek notion of scientific explanation consisted of an accurate description of a phenomenon, the analysis of its main characteristics, and the relation of these to a series of universal truths, principles or laws already put forward. What distinguished Greek science from modern science—apart from its limited numerical notation—was the nature of these universal truths and the ways in which they were recognized with certainty. Greek thinkers of course disagreed among themselves about these universals. Plato and his followers believed that they were Ideas, and the aim of their 'science' was to trace the perfect world of these Ideas of which the material world was but a shadowy and imperfect model. Aristotle, once Plato's pupil, denied the separation between Form (or Idea) and Matter (or Substance). He and his followers believed that Forms could be analyzed from the perceptions of the senses—hence their interest in classification and description. Explanation was thus achieved by fitting facts into an existing (*a priori*) scheme of Forms. This required reason: analysis of the facts, resulting in organization and interpretation.

In the **Physics**, Aristotle presents the known facts about motion and change. He then considers the meaning logically to be given to the idea of motion. Motion is to be explained by the tendency of the thing in question to get to its natural place of rest. Thus fire naturally goes up; the resting-place of earth, on the other hand, is down. This means that explanation is given by reference to the end of the motion—where an object goes to and why it is going—rather than how it is getting there. Causes are explained in terms of the goals achieved: an animal grows because it is to become an adult; a seed germinates because it is to become a plant. These are Aristotle's **final causes**. The tendency of an object to move, grow or change—of a ball to roll downhill, for example—is a **formal cause**. Such a cause may not in fact be operative if the ball is held on the slope, but it remains a potential fact. In contrast to this definition of causality in terms of future goals or possibilities, causation in modern science is defined with reference to Aristotle's **efficient causes** acting on passive materials or **material causes**. The fundamental methods of explanation are thus totally different.

This preoccupation with the final end of processes rather than with the processes themselves led Arististotle and his successors to concern themselves with logical connections—what is the essence of an entity? What is the potentiality of an object or activity? The categories of mass and force in the context of space and time simply did not arise in fundamental explanation or knowledge—*scientia*.

The concern for purpose was linked to the Idea of reason governing the whole universe. It was linked to the orderly arrangement of everything in a cosmic hierarchy, and to the ultimate rationality and purpose of God. For the Medieval schoolmen, the underlying reasonableness of the universe was a basic assumption; so was the penetration of all things by the hidden purpose of the Creator. God was understood as the First Cause not only in the sense of the originator of the cosmos, but also as the underlying Will within each lesser purpose. Thus God's will and purpose in creation constituted the final explanation of all things; their existence, motion or change.

The thought of both Plato and Aristotle assumed the necessity of God, whether conceived of as the ultimate Idea or Form of the Good in Plato's case, or as the Prime Mover in the thought of Aristotle. Thanks to the genius of St. Augustine of Hippo and later St. Thomas Aquinas, the Medieval Church was able to reconcile these abstractions with the Christian notion of the God and Father of Jesus Christ. Because of this, the thought of the Middle Ages was predominantly deductive. It was inevitable that men of faith and reason should proceed from general principles and universal truths to particular examples of those truths, rather than start from particular pieces of evidence and then form generalizations about them. It was the very considerable achievement of the early scientists to take the latter path, and they may surely be forgiven if they were reluctant to pursue this course at every point in their thinking. The dominance of Aristotelian categories of explanation made it enormously difficult to keep clear the alternative types of description, and at the same time to hold to the accepted patterns of reason and

figure 1 **Ptolemy's System**

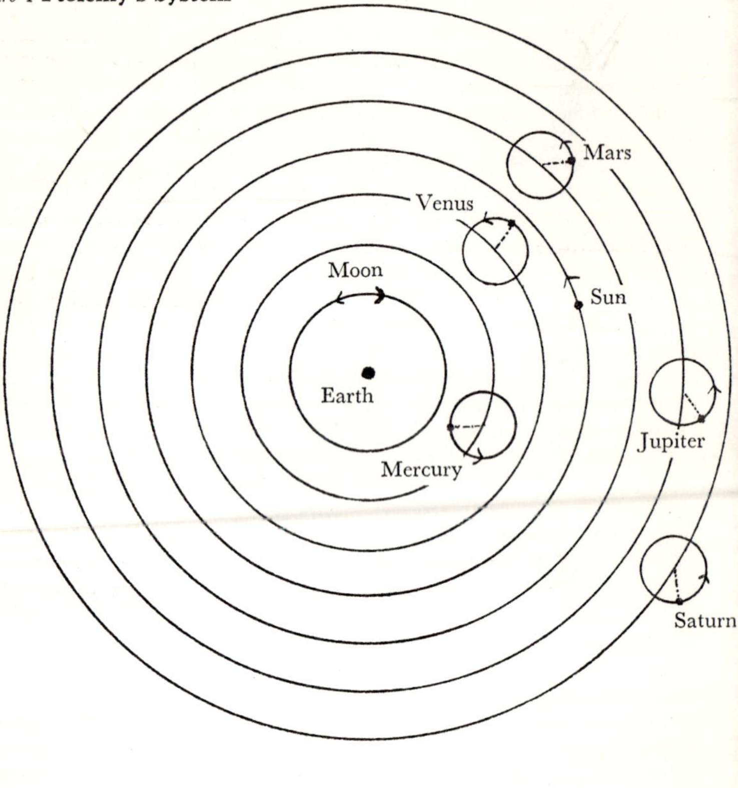

faith. Galileo was obliged in his scientific work to set aside all notions of final causes, though he succeeded to his dying day in holding his faith in God as First Cause and Father of all.

Another central feature of the Medieval scheme was its cosmology or model of the universe. Here the picture developed by Ptolemy of Alexandria (c. A.D. 100–168) was adopted. Ptolemy's model had the earth at the centre and the sun and other known planets revolving around it. The planets moved in small circles or epicycles about points which themselves moved in fixed circles around the earth. In this way Ptolemy explained the available observations of the sun and planets, and preserved the deep-seated religious conviction of his time that heavenly bodies moved in perfect (that is, circular) ways (*figure* 1).

On the face of it this picture was realistic enough. The crudities of astronomical measurements did little to bring it into doubt until Copernicus performed his calculations and Tycho Brahe improved the system of measurement. Man was given a unique position at the centre of things by this model, and the Medievals believed that it was supported by the scriptures as well as the authority of the Church. Although the divine realm was more perfect than the observable world, and at a different level of being, it could react upon this system at every point. The divine laws could be discerned in the plan and order of the cosmos, which was a kind of allegory of the moral and spiritual order. The system was essentially fixed, and was naturally regarded as a place on which each human soul could work out its eternal destiny, as St. Augustine had taught. To use another metaphor, the created order was a stage on which the age-long drama between good and evil was played out by individual men or nations, and from which mankind would finally be raised and judged for heaven or hell. The Medieval Mystery plays, Dante's **Divine Comedy** and the works of Shakespeare assumed this picture which the new science was to overthrow.

3
New Experiments, Mathematics and Theories

The major philosophical system of the late Middle Ages owed its influence and acceptance in Christendom to the Dominicans or Black Friars, founded in 1215, by St. Dominic (1170–1221). He was a professional theologian and cathedral dignitary when he established his order of preachers to spread true doctrines and confound heresy. The most illustrious of the Order were men like Albert the Great (c. 1200–80) and Thomas Aquinas.

The rise of experimental science, on the other hand, owes much to the work of members of the second religious order founded in the thirteenth century, the Franciscans or Grey Friars (1209). St. Francis himself was not a man of learning, and intended that his Poor Brothers—as he called them—should lead a simple, joyful life, preaching to ordinary people the simplicities of the Gospel. The wandering Friars soon found that they had to contend with all kinds of heresies and turned to study and scholarship in order the better to preach against them. Nevertheless, St. Francis' very positive attitude towards the created order—vividly expressed in the **Canticle of the Sun** which is attributed to him—encouraged his followers to interest themselves in the creation and its marvels. Among the early Franciscans who expressed this in both teaching and experiment were Robert Grosseteste of Oxford, who performed experiments with mirrors; John of Peckham (c. 1220–92), who also considered optics in the light of Arabic writings; and Roger Bacon (c. 1214–94).

Bacon, unlike Grosseteste and Peckham, was an ordinary friar who reached no high office. Born in Somerset, he studied at Oxford and Paris and in 1250 was lecturing at Oxford. After losing his wealth, he joined the Franciscans, but before long tried to take up his scientific activity again. After a long period in which he was forbidden to write by his superiors, a friend became Pope (Clement IV) and arranged for him to continue his scientific studies. Two years later, in 1268, Bacon presented the

Pope with his **Opus Majus**, a compendium of scientific ideas of the time.

Bacon read and wrote more than he performed experiments, and without doubt allowed his imagination to run riot at times—proposing mechanically propelled carriages, flying machines and ships, possible uses of gunpowder and lenses in spectacles and telescopes. In practice he must have done some experiments, because he describes the laws of optics—reflection and refraction—and was able to explain lens magnification. In the **Opus Majus**, he argued the importance of mathematics as a basis of knowledge, and insisted that scientific knowledge could only be acquired by experiment. These two key ideas were fundamental to the progress of the scientific revolution. His story also illustrates the prejudice against the pursuit of science in the Church at the time.

Other Franciscans who deserve notice include Oresme, Bishop of Lisieux (1323–82), a mathematician and economist as well as an administrator and theologian, who wrote a tract on currency using vulgar fractions—$\frac{2}{3}$, $\frac{3}{8}$, etc.—and challenged the Aristotelian view that the earth was fixed. Cardinal Nicholas of Cusa (1401–64)—another mathematician and influential thinker—likewise rejected the traditional theories of astronomy, asserting that the earth moved like the other stars. These names at least indicate the positive attitude which was sometimes adopted within the Church towards experiment, mathematics and new theories of different kinds.

In the fifteenth century, laymen as well as religious may be numbered among the pioneers. Leonardo da Vinci (1452–1519), the illegitimate son of a Florentine lawyer and a peasant girl, was not only a great artist and sculptor, but architect, engineer, philospher and experimenter as well. He did practical work—somewhat slowly and clumsily—in optics, mechanics and hydraulics. His notebooks are full of plans and designs for guns, mortars, helicopters and flying machines. He hit upon the explanation of the dim light given off by the dark part of the moon's disc at new moon, recognizing it as 'earthshine'—light reflected from the earth. He anticipated Harvey's discovery of the circulation of the blood, and suggested correctly various functions of the blood in the human body. He believed acceleration was produced by external, not internal, impetus, as the Arsitotelians held, and thought the earth was 'another star' possibly revolving round the sun. All these ideas were put forward without experimental verification, but based on existing data and a gigantic imagination. He also suggested that the universe might be controlled by mechanical laws—repeating the speculations of Democritus and Anaximander—and anticipating the better-founded laws of Newton.

The fact that Leonardo could think so freely is itself testimony to the atmosphere of the fifteenth century in which the next great phase of science began. Like Roger Bacon, Leonardo insisted that science should be based on empirical observation; it may use mathematics to 'discuss' the observed data, and ought to end with a crucial experiment to test whatever theory is thereby developed. This in outline has remained the ground-plan of modern science.

One of the first men to use the improved mathematics to profound effect was the Polish Churchman Micola Koppernigk, better known as Nicolaus Copernicus (1473–1543). After studying at Universities in Cracow, Bologna, Ferrara and finally at the great centre of mathematical study and experimentalism in Padua, Copernicus entered the ministry of the Church and attained high office as an administrator, diplomat and government adviser. He was brought up at school and university to accept the 'official' astronomy of Ptolemy, but like a number of his contemporaries had doubts about its validity. Reviewing other possible explanations of the relation between the earth, other planets and the sun set out in antiquity, Copernicus examined the available—and ancient—mathematical data with great care. In his work **De Revolutionibus Orbium Coelestium**, published in 1543, the year of his death, he proposed an alternative model to provide a more satisfactory explanation of the facts—a model in which the earth revolved on its own axis, and—with the other planets then known—revolved around the sun. This proposal, supported by the mathematical skills available to Copernicus and of which he was a master, indicates a vital fact in the development of science—that observed data need adequate interpretation; bare facts, in mathematical or other form, have to be placed in a satisfactory conceptual framework for understanding to advance.

The new theory at once presented the small band of practical astronomers with something to test for confirmation or disproof, though few of them took it very seriously to start with. The model made further experiments desirable and possible. Such is the method of this pattern of human understanding. The same processes may be discerned in the work of Versalius on human anatomy which was published in the same year as that of Copernicus. It is to be found in every stage of scientific thought since then.

Perhaps the greatest of the sixteenth century astronomers was the Dane, Tycho Brahe (1546–1601). He was not a notable mathematician or theorist—in fact he subscribed to Ptolemy's model of the universe, and wanted to disprove that of Copernicus—but he was a first-rate observer and practical scientist. He designed new astronomical instruments, on a much larger scale than before with a corresponding improvement in the accuracy of his measurements. He introduced the procedure of taking large numbers of observations and averaging the result. By these devices he raised the standard of accuracy of astronomical observation out of all recognition and made available new, reliable and detailed data which was to prove decisive in the theoretical work of his brilliant assistant, Johann Kepler. It was Brahe who first observed a *nova*, or new star, which appeared suddenly on 11 November 1572, and then slowly faded. The idea of such change in the panoply of outer space was one more nail in the coffin of the Aristotelian doctrine of its changelessness and perfection.

Some of the possible implications of the new science received the speculative attention of another figure of this time—the Dominican, Giordano Bruno (1547–1600). Although under suspicion from his superiors as a heretic on the subjects of transubstantiation and the conception of the Virgin Mary, Bruno continued his imaginative adventures in the realms of cosmology. He denied the infinity of God, and suggested that there might be endless worlds similar to the earth in other solar systems. He thus displaced both the sun physically and God theologically from the centre of the universe. In Bruno's scheme no place could be found for heaven or hell, and pantheism reigned. Inevitably the Church reacted through the Inquisition. Bruno was arrested when he returned to Italy from London in 1593, imprisoned for seven years, brought to trial and burnt at the stake. There is no excuse for such brutality on the part of the Body of Christ, but given the convictions of the time, it is hard to see how the Church could have tolerated such disastrous theology. Bruno, who was never a practical scientist, fell into the fatal error of confusing the substance of his theology with the new cosmology, and confirmed the worst fears of conservative thinkers about the new scientific discoveries.

Just as Luther had laughed with characteristic rudeness at Copernicus as 'the fool who would overturn the whole science of astronomy', so others later in the sixteenth century looked with suspicion on the small number of men who entertained new hypotheses about the cosmos, or the human body. The famous trial of Galileo Galilei in 1632 was an expression of that reaction to the still more fundamental attack on the Medieval scheme contained in his popular writings on astronomy and mechanics. The attitude of many in the Church of England to the Royal Society in London in the early years of its foundation after 1662, was the same, although the quiet progress of the scientific revolution in England was unimpeded by ecclesiastical opposition.

Looking back on the affair of Giordano Bruno, it may seem that the whole conflict was unnecessary, yet the leaders of the Church had to defend their flocks, and Bruno's theology was obviously unacceptable. It is hardly to be expected that pious and intelligent men should agree with an upstart and discredited monk that their reasoning, teaching and patterns of conviction were absurd. The problem touched men very deeply, and it was perhaps inevitable that it should have broken out in violence at some point. Nevertheless it seems that scientific progress was delayed less by ecclesiastical opposition than by ignorance of the new ideas and indifference to them. It was theology which was retarded by the conflict as much as scientific understanding. That lesson from Bruno's fate has had to be relearnt many times since then.

4
From Galileo to Newton

In the history of science it is now seen that the towering figure who brought about the final overthrow of the old cosmology was the Italian, Galileo Galilei (1564–1642). The son of a nobleman of Pisa who was a musician and mathematician, Galileo was educated in the Aristotelian tradition. He declined to enter a religious order, going to the University of Pisa instead to study medicine. There he transferred his interests to mathematics and science. At the age of 25, he was appointed to a lectureship, but after only two years—and many quarrels with his traditionalist colleagues—he took up the Chair of Mathematics at the University of Padua, the most advanced centre of mathematical study at the time, and a university noted for its freedom of thought.

While still at Pisa, Galileo had begun to question the conventional doctrines of Aristotle on the subject of motion. To test his opinions he performed various experiments with weights, although according to Professor Butterfield, there is little evidence that he dropped balls of different weight from the famous leaning tower. These experiments led Galileo to the study of mechanics and the formulation of his well-known laws of acceleration: the speed of a falling body at any point in its descent is proportional to the time which has elapsed since it was released, and this speed of fall increases uniformly with the time. To test this hypothesis Galileo had to overcome the problem of timing. He did this by using the weight of water from a water clock, released as a polished steel ball was allowed to roll down a gently grooved slope. This experiment showed clearly that a force (gravity) had the effect of **changing** motion, producing acceleration. He could also show that a body on which no force acted would continue to move at a uniform speed by allowing his polished balls to roll on to a horizontal plane and continuing to time them.

In this way, Galileo confirmed the conjectures of Leonardo da Vinci and of his contemporary, the philosopher and engineer René Descartes. Galileo's 'beautiful experiment', and the law he confirmed by it, went far to revolutionize the whole of mechanics.

Galileo also studied the motion of projectiles and showed that the normal path of a body thrown forward would be a parabola. He studied the motion of pendula—timing the swing of a roof lamp in the cathedral in Pisa by the beats of his heart—and discovered that the time of swing of a pendulum of any weight depended only on the length of the suspension and not on the size or weight of the swinging mass. Later, he realised the possibilities of the pendulum for a clock mechanism, though he never made such a device. He also considered the phenomenon of 'centrifugal force', showing that it is proportional to the square of the speed of the moving body, and inversely proportional to the diameter of the circle which the body follows. Newton was to make use of this formula in his calculations of the force of gravity between the sun and its planets to maintain them in a circular orbit.

Galileo's other investigations included the study of density, thermometry, the strength of materials, the grinding of lenses and optics, and the use of the telescope in astronomy. He is perhaps best known as the man who first 'turned a telescope on the heavens'. The story is told of a professor of philosophy at Padua, of an Aristotelian persuasion, who refused Galileo's invitation to look through his telescope on the grounds that it could not possibly tell him anything he did not already know!

Important as Galileo's astronomy was, the fundamental achievement of this remarkable man lay in the field of mechanics. It was he who made the decisive breakthrough to the new pattern of explanation, replacing the Aristotelian categories of motion with new ones of force and weight which could be measured quantitatively in time and space and represented mathematically through algebra and trigonometry.

Galileo's outspoken opposition to the Aristotelian system expressed in his writings on astronomy caused concern to the ecclesiastical authorities, not least because his written publications were in Italian rather than Latin and were received by a large audience. After returning to Pisa under the patronage of the Grand Duke in 1610, he was warned to avoid making statements having a theological implication. His work continued, and in 1632 he published the **Dialogue on the Two Chief Systems of the World**, a popular discussion of the contrast between the old and new cosmologies—in which he was not wholly fair to the old system. Thereafter he was brought before the Inquisition in Rome and forced to recant all his views as contrary to the authoritative teaching of the Catholic Church. During the fascinating trial, Cardinal Bellarmine suggested that the new cosmology might be used as a convenient calculating hypothesis provided that it was not accepted as a true representation of reality, but although Galileo was prepared to consider this face-saving proposal, the court required him to recant the opinions he had clearly set out in the **Dialogue** and condemned him to silence. In spite of the ban, this irrepressible controversialist managed to have his greatest work **Mathematical Discourses and Dialogues concerning Two New Sciences** published in Leiden in 1638.

Important as Galileo's experimental work and publications are, it is now recognized that his achievements—such as the explanation of motion—have to be understood in the context of a development which goes back to Nicole Oresme in the fourteenth century. Like other men of genius, his concepts were the most successful in a long sequence of groping. At the same time

Galileo was dogmatic in his own way, and certainly did not appreciate all the consequences which his new explanations were to have in the realm of theology.

Contemporary with Galileo was the German mathematical astronomer Johann Kepler (1571–1630). The two men corresponded from time to time, but their intellectual and actual paths never crossed. Unlike Galileo, Kepler was a Protestant, a highly abstruse writer, and a man absorbed in mathematics and their harmony rather than in practical observation or mechanics. Using the observational data he inherited from Tycho Brahe, Kepler was able to construct a modified model of the Copernican system. He showed, with the aid of new mathematical tools like logarithms and the improved trigonometry, that the planets moved in elliptical orbits with the sun at one focus. His Laws of Planetry Motion are the first example in modern science of the result of a theory being sought to fit the observed facts, taking account of probable experimental errors. Kepler sought universal generalizations from his data rather than piecemeal explanations, and put on one side all other considerations save that of the possibility of providing a total, systematic explanation or **metaphysic**. As a result, Kepler filled in many of the gaps and inadequacies of the Copernican system, presenting a far more radical alternative model in his new cosmology of dynamic motion and change. In short he provided a totally new world-view.

Another contemporary of Galileo was René Descartes (1596–1650). Born at Tours and educated by Jesuits, Descartes turned to a career in mathematics and engineering, to which from the age of 25 he devoted his life. In his first book **Le Monde**, the publication of which was delayed by news of Galileo's trial and condemnation, Descartes made the first attempt to explain the universe in terms of mechanics. It was not an experimental treatise, and contained hard words about Galileo's 'philosophy of falling bodies'. In his later works, **Discours de la Méthode** (1637), and **Principia Philosophicae** (1644), Descartes set out a new natural philosophy, first by making a new classification of the qualities of substances. These, he said, were either **primary**

—such as extension in space or motion; or **secondary**—such as hardness, sweetness or roughness. On this analysis, Descartes went on to deduce specific laws of inertia: that all bodies continue in motion in a straight line unless they are acted upon in some way; and the law which states how the motion was distributed between two bodies on impact. Interestingly, Descartes and his followers could only think of force and motion in terms of contact between objects or particles. They were unable to imagine the idea that a force (such as gravity) could act on a body from a distance. Descartes' writings laid the foundations for subsequent corpuscular theories of matter and energy.

From the point of view of the interaction of science and religion, Descartes' contribution was extremely important. He reasserted in a new and powerful way the Hellenistic distinction between matter (extended in space) and mind (unextended 'thinking substance'—totally unrelated to the matter of the external world) Everything between matter and mind—animals, plants, even the human body—was for him part of the mechanical world. The mind, which he located in the head, was the realm of thoughts, feelings, desires, ideas—and especially of purposes and final causes. Thus Descartes was able to reconcile (after a fashion) the theology of Aquinas with the new mechanistic ideas of science. In the process, a drastic separation was made between mind and matter; the observer and that which is observed. It was this which made room for the development of the Newtonian world-machine, literally conceived as a billiard-ball universe with God attached. This philosophical assumption—the 'Cartesian Split'—has produced many of the major problems of philosophy and of the relation between science and religion. How does science describe the world? What is the relation between mind and matter? Is there any place for purpose in the material universe? If such purpose cannot be discerned by scientific methods, is there any objective way of discovering it? What do we mean by the human mind anyway?

When Christian thought attempted to come to terms with this way of looking at the world in the Age of Reason which was to follow, it came near to losing itself in Rationalism. In the thought of the Deists, God was removed to a distant place above the firmament of heaven where he simply watched the workings of the mechanical laws he had made. Some came later to believe that there was no rational evidence for God at all—Scepticism and Atheism emerged, especially among the *philosophes* in France. Even John Milton, in his heroic poems, reinterpreted the fall of man from Paradise in accordance with the new stress on Reason. So Adam says to Eve in **Paradise Lost**:

But God left free the will, for what obeys
Reason, is free, and Reason he made right.

Paradise Lost *IX*, 351.

The Fall itself becomes a psychological and 'rational' affair, not, as the biblical story suggests, the result of disobedience. The realm of the mind, as a separate realm from that of the everyday world of objects, becomes the realm of religion.

Many modern problems in the relation of religion to the ordinary world, which is also described by science, may be attributed to the hangover from the philosophy of Descartes. Certainly the relegation of religion to the mind or the feelings—to private life rather than to public action in a technological world—still mirrors Christianity's subservience to this philosophy.

The rejection of scholasticism brought losses as well as gains. Instead of the vibrant world of the Middle Ages, full of life, colour, sound and meaning, the new philosophy introduced a world of Nature, which Professor E. A. Burtt has described as follows:

The really important world outside was a world,
hard, cold, colourless, silent, and dead—a
world of quantity, a world of mathematically
computable motions in mechanical regularity.

The Metaphysical Foundations of Modern Science *p.* 237.

But Descartes' philosophy succeeded. As Professor Basil Willey points out, the men of the Renaissance began to look for a new life-orientation, a new control over **things**: already the satisfaction of science and the progress of technology had created this new climate of opinion. What happened in the scientific revolution was a transference of interests from the total being of things to the abstractions of science and its **use**. It was against what they believed to be the overstress on the abstractions that men of faith were constrained to protest. Men like John Donne and Jonathan Swift attacked the new sciences, fearing what they would lead to. As the German poet Schiller wrote:

Ah from that warm and living vision,
Only the shadow remains behind today . . .
Like the dead strokes of a pendulum-clock,
Nature, bereft of all her Divinities,
Slavishly serves the Law of Gravitation.

Christianity had to endure the dark night of Rationalism and the excesses of the Evangelical Revival before a balanced attempt could be made by E. Schleiermacher (1768–1834) to assert a new totality of faith. Meanwhile the new philosophy of Descartes carried the day in progressive, intellectual circles. The conviction grew that all natural phenomena could be reduced to the motion of matter. Dynamics became the basis of science; corpuscles, the models of matter. And as Galileo had provided the concepts and mathematical methods by which the motion of bodies could be described, so that great genius Newton, born in the year Galileo died, showed by experiment and conjecture that this was how the world could be understood. Later, Alexander Pope wrote for Newton's epitaph:

Nature, and Nature's laws lay hid in night:
God said 'Let Newton be!' and all was light.

Isaac Newton (1642–1727) was born on Christmas Day at Woolsthorpe, near Grantham, in Lincolnshire. After an unadventurous schooling in Grantham he was sent to Trinity College, Cambridge in June 1661. There he discovered an interest in mathematics and science. He quickly mastered the available material, going on to consider a number of outstanding problems and finding solutions for them. Independently of Leibnitz, he invented a calculus and developed other mathematical techniques like the method of approximating an infinite series. He became interested in the phenomena of light, in telescopes, and in gravity—a force 'extending to the orbit of the moon'. Discarding Descartes' theory of vortices, Newton developed his idea of gravity on the evidence of direct experience. Using Kepler's Rule he deduced the law which states that the forces which keep the planets in their orbits must be in proportion to the inverse square of the distance from the planet to the centre of the orbit.

Newton returned to Cambridge (after absence during outbreaks of the bubonic plague) and became Professor of Mathematics in 1669. There he carried out experiments in optics and chemistry as well as mathematics. After a period of fruitless interest in Alchemy—Newton had a strange streak of mysticism in his temperament—his friend Edmund Halley, the astronomer, persuaded him to return to mathematics and encouraged him to produce the **Principia**. Later, Newton sat as a Member of Parliament, became involved in the political circle, had a nervous breakdown which ended his scientific career, and, as the Master of the Royal Mint from 1698, received the honours of his contemporaries in science and society. Strange to say, only now are many of Newton's papers being edited for publication. They confirm the view that he was indeed an intellectual giant.

Newton's greatness lay in the power of his intellect and imagination, and in the unique combination of experiment, mathematics and theory in his science. He believed that the task of the scientist was to describe, or, in the face of inadequate evidence or means of interpretation, to admit to ignorance. Like Descartes and Galileo, Newton was a literalist, believing that scientific description was a literal representation of objective reality—a view which was to be challenged by Kant and which is now under careful reconsideration. In his developed picture of the world as a mechanism of objects and particles, Newton presented what he thought was a complete account of its nature. Thus a technique of explanation was turned into a total system—a metaphysic—and the only possible one at that.

In spite of this, Newton still found a place for God and the human soul. He was a devout Christian all his life, and devoted as much time to theology as to science. He argued that the order and beauty of nature require faith in an 'intelligent, incorporeal, living being . . . a Designer'. The very order of the natural world was a sign of divine benevolence. In Newton's scheme the continuing function of God was the adjustment of the solar system. He thought that there were no scientific explanations for the pattern of the planets, or for certain continuing irregularities in their motion. God somehow kept the stars from falling together, and was identified with the notion of absolute space and time. Thus the 'God of the gaps' was introduced, and the idea of God's immanence in nature was lost by those who accepted the Newtonian scheme.

The scientific and theological inadequacies of Newton's scheme were soon to be exposed. Laplace declared he had no need of the hypothesis of God in his explanation of planetary motion, and Leibniz pointed out that a perfect God would not have created an imperfect world in constant need of correction. The path to Deism—the notion of a Divine clockmaker who makes the universe then sits afar to watch it go; and the path to scepticism or Atheism—the notion that there is no evidence to support the idea of Deity at all—are here made plain. For all his very sincere piety, the assumptions implicit in Newton's world-machine were to overrun the popular idea of God, divorced as it had become from the revelation of the Bible. In the next decades, theology could only meet the new metaphysics by adopting them, as large sections of Protestantism did; or by reacting against them by upholding the thought of St. Thomas Aquinas—the path followed by the Roman Catholic Church from the time of the counter-reformation until very recently. Meanwhile, in France particularly, scientific method and theory came to be regarded as providing total explanations of reality. Scientism and humanism were born.

Finally, the work of the scientists of the sixteenth and seventeenth centuries contributed greatly to the new Renaissance estimate of man. Galileo's defiance of the injunction of his Church was more than obstinacy; it was an expression of faith in human reason. Descartes' powerful natural philosophy gave pride of place to the human mind and its powers of reasoning, not least in mathematics. Newton's world-picture was one of a moral, ordered system in which man was distinguished by his reason. This alone marked him as different from the world-machine; and yet this associated him with it, since man, like the machine, is reasonable and orderly.

Great genius as he was, Isaac Newton was very much a child of his time. His age was an age of reason and formal organization, especially among the leisured classes. Formal gardens and architecture went hand in hand with formal courtesies and art, and even with reasonable politics and toleration. The implied rational notion of knowledge in the Newtonian scheme was taken up and made explicit by the influential philosopher John Locke (1632–1704):

> *I say then, that we have knowledge of our own existence by intuition; of the existence of God by demonstration; and of other things by sensation.*
>
> **Essay Concerning Human Understanding**, *IV, ch. 3*

Taking scientific knowledge as his standard, Locke looked for a science of human nature and society, and greatly encouraged the belief that reason must prevail in every field of human endeavour. Sigmund Freud and Auguste Comte were later to provide the foundations for scientific psychology and sociology respectively, but even in Newton's day, the philosophers Hobbes and Spinoza sought to reduce the human mind and society to rational terms.

From the time of this Age of Reason, or Enlightenment, it has been popular to maintain that man was the reasoning and reasonable being. In this he was thought to be like God—if people felt there was room left for God in such an ordered, explainable world. The implications of this view for the Church, Catholic or Protestant, brought up on the Medieval interpretation of the drama of sin and salvation were considerable. They are still with us in education, counselling, preaching and the administration of the law. They pose the questions: Does wayward man need salvation or better education? Should criminals be reformed or punished? Do heathens need enlightenment or conversion? Is man made in the image of God—or God in the image of man?

5

Topics, Questions and Sources

1
Find out what you can about: *a* the astrolabe, *b* an alchemist of Chaucer's time, *c* the sources of Chaucer's tales.

2
What were the teachings of St. Thomas Aquinas about: *a* the relation between reason and revelation, *b* the idea of God as the First Cause, *c* the purpose of the universe?

3
Make an exhibition showing the interests and achievements of one of the following: Leonardo da Vinci, Galileo Galilei, Isaac Newton, Nicolaus Copernicus, Nicholas of Cusa.

4
Make a collection of hymns, poems, pictures and music to illustrate the emphasis on human reason and the orderliness of creation in the period 1660–1714.

5
Dramatize the trial of Galileo **or** read and discuss Brecht's play **The Life of Galileo**, *Methuen* 1963.

For Discussion:

6
Why did natural science arise in a Christian civilization, and not in any other?

7
Was the Church right to condemn Giordano Bruno for his suggestion that there might be other planets similar to the earth in other solar systems?

8
Can a scientific experiment prove a theory to be true?

9
Is it still possible to explain things by the idea of purpose?

Sources:

General—
The Scientific Revolution A. R. Hall *Longmans* 1962.
The Origins of Modern Science H. Butterfield *Bell* 1950.
The History and Philosophy of Science L. W. H. Hull *Longmans* 1959.

Astronomy and Mechanics—
The Growth of Physical Science J. Jeans *C.U.P.* 1948.
The Fabric of the Heavens S. Toulmin & J. Goodfield *Penguin* 1965.

Religion and Philosophy in the Scientific Revolution—
The Seventeenth Century Background B. Willey *Penguin* 1962.
Science and the Human Imagination M. B. Hesse *S.C.M.* 1954.
Issues in Science and Religion (Part I) I. G. Barbour *S.C.M.* 1967.
Aquinas F. C. Coppleston *Penguin* 1959.

On Individual Scientists—
Copernicus, the Founder of Modern Astronomy A. Armitage *Yoseloff* 1938.
The Watershed (Kepler) A. Koestler *Heinemann* 1961.

The Crime of Galileo G. de Santillana *Chicago* 1953.
Isaac Newton L. T. More *Dover Books PB* 1934.
A Discourse on Method René Descartes *Penguin* 1964.
Note also the **Men of Genius** Series *Nelson* and the **Phoenix** Series *University of Chicago*.

B
New Explanations

1
The Sciences of Life: Biology, Anatomy and Physiology

About the year 600 B.C. on the Island of Cos, there arose what is thought to have been the first school of medicine. From that school came a number of books written by some of its leaders, including a man who is sometimes called the father of medicine, Hippocrates. These books give a clear indication of the kind of observation made at the time, and of the methods used to make them. One volume, **On the Sacred Disease**, sets out the important scientific dictum that a natural explanation is to be sought for all observable events.

The bulk of early Greek biological writing concerned outward appearances of marine creatures and land animals, but in 500 B.C. Alcmaeon made dissections and described the nerves of the eye and the tube connecting the nose to the ear. He began the study of embryology—the development of young animals—incubating eggs and observing features of the growth of the chick.

It is from the Hippocratic school that the doctrine of the **Four Humours** came. All living bodies were thought to be made up of four humours: Blood (*sanguis*), Yellow Bile (*cholera*), Black

Bile (*melancholia*) and Phlegm (*pituita*). These humours were related to the four elements of which matter was thought to be composed: earth, air, fire and water. Health depended upon the correct blending of the humours. An inbalance was thought to lead to disease which could be sanguine, choleric, melancholic or phlegmatic. Each humour was associated with a particular organ; blood with the liver, melancholy with the spleen, phlegm with the lungs, and choler with the gall bladder. This theory was to make a very deep impression for two thousand years, particularly in the development of the psychology of personality. Its traces are to be found in the plays of Shakespeare, and in the loom of our language to the present day.

Aristotle (384–322 B.C.) was the great biologist of antiquity, perhaps of all time. Of his books which have been preserved, four major works deal with biological matters—**On the Soul** (*De Anima*) dealing with the principle of life, **Histories about Animals** (*Historia Animalium*), **On the Generation of Animals** (*De Generationem Animalium*), and **On the Parts of Animals** (*De Partibus Animalium*); together with a number of lesser works. These books deal with observations and theories, as all scientific works must do. Among the creatures Aristotle studied and classified are the catfish, the electric torpedo fish and the angler fish; the octopus and cuttlefish; the whale, porpoise and dolphin; the placental dogfish and the honey bee. To him we owe the basis of classification of living things—the idea of *genus* and *species*—and the first attempt to state the difference between living and non-living things in terms of *psyche*. This he further differentiated into vegetative, sensible (animal) and rational (human) psyche, or soul. Once again these ideas persisted through to the Middle Ages and leave their traces in contemporary forms of speech—vegetative life, sensual feelings and rationalism. In spite of this differentiation, in his later works Aristotle inclined to the modern view that man is akin to the lower creatures in sharing 'life' with them, and he thought that there was no fundamental distinction between life, soul or mind.

Aristotle's main interest lay in the study of animals—zoology.

His successor at the Lyceum—the name of his school or academy in Athens—Theophrastus, laid the foundations for the other traditional branch of biology—botany. From Theophrastus come two major works on botany, one of which is the famous **History of Plants**, a collection of notebooks and stories about plants and their uses. Like his former master, Theophrastus gave names to parts of the organisms he studied, and showed particular interest in the process of development. He distinguished various modes of reproduction, though he held to a belief in spontaneous generation—that sexual reproduction was unnecessary and that living forms could emerge from inanimate matter.

The important scientific school at Alexandria added to the growing store of biological and anatomical knowledge through the work of men like Herophilus on the brain and the circulatory system of the blood (c. 280 B.C.), Erasistratus (c. 270 B.C.) on the nervous system of man and higher animals, and Crateuas (c. 100 B.C.) who gathered and illustrated herbs in a realistic way for the first time. In the first century A.D., Dioscorides, a military doctor and naturalist, and Pliny, the Roman civil servant and collector of tales and ancient information, each made their contribution and were influential in the following centuries.

Most important of the later ancients is the medical writer Galen. Born at Pergamum in Asia Minor in A.D. 130, Galen attended the rival school to Alexandria in his home town. He travelled widely, eventually became physician to the Emperor Marcus Aurelius and wrote prolifically on every aspect of medicine. He studied anatomy and performed dissections on several kinds of animals, though never on the human body. He investigated the function of various structures and organs such as the spinal cord, the heart and lungs. He may therefore be called the first physiologist. Although he was an able investigator, he made a number of mistakes, and as often happens, his errors survived his wisdom in the abridgements of his vast writings which were passed on to the future. Such was Galen's authority that his ideas were still being taught in the university medical schools of the seventeenth century. This was due also to the religious ideas which permeated his writings. Although he was not a Christian, Galen held many of the basic Hebrew-Christian ideas about God, and maintained that every organ of the body was specifically designed by God for its purpose in the most perfect form—a notion which readily accorded with Christian piety both of the Dark Ages and of the Medieval synthesis with Aristotle's philosophy.

Little is known of any significant development in the fields of biology or anatomy during the next thousand years and more. The Dominican Albertus Magnus (Albert the Great, 1206–80), produced an encylopaedia of science based on Aristotle including many comments of his own. He also published a book **On Animals** discussing the blood supply of embryos and another **On Plants** with detailed descriptions of tree and plant structure —perhaps the best natural history of the Middle Ages. These were a vast improvement on the quaint Medieval bestiaries.

Medicine did not flourish so well because of the steady opposition of the Church to human dissection—even of corpses. This opposition was founded on two premises: first, the view that since God had made each organ for a specific purpose, it was wrong to interfere with it unless—in the case of an amputation, for example—the law of preserving life could reasonably overrule other considerations. Dissection for scientific purposes could not be justified by this kind of moralistic thinking, and in the light of existing knowledge and the absence of anaesthetics this was perhaps as well. Secondly, the doctrine of the resurrection of the body in the Church's creed was taken to mean that on the day of resurrection, the mortal body would be somehow raised up in a literal fashion, and therefore it was unthinkable for anyone to tamper with it, even after death. At the University of Bologna, however, the medical faculty, which was associated with the faculty of law, undertook *post mortem* examinations of victims of murder, and human anatomy revived under men like Mondino (c. 1270–1326). But for the most part, medical education consisted of book-learning on the lines of Galen's knowledge, and the physician was held in far higher esteem than the surgeon.

A major innovation occurred when Vesalius began to give anatomy lectures at Padua in 1537, illustrating human anatomy himself with the aid of specimens before his class. His classic treatise **De Humani Corporis Fabrica**, published in 1543 in the same year as Copernicus' **De Revolutionibus Orbium Coelestrium**, was the first modern textbook of anatomy.

The religion of the ancients, tempered with their rational philosophy, encouraged their investigations into the realm of nature. The needs of the sick stimulated early botany and human biology. Paradoxically, in spite of the teaching of Jesus, the influence of Christianity inhibited further development of these sciences—indeed it may even have put the clock back. Many teachers in the early Church were deeply suspicious of classical learning, Christians were forbidden to teach or send their children to classical secular schools. Early Western Fathers like Tertullian (c. A.D. 200) were adamant about this, so too was St. Augustine. A synthesis of Christian thought and ancient wisdom was achieved in the Christian East, centred first on Alexandria and, later, on Constantinople (Byzantium), but by then the two parts of the Church were growing apart, and much of the learning of the Byzantine Empire was lost with the coming of the Turks.

The Arabic scholars, anxious to claim all wisdom for Mohammed, collected and preserved much of the ancient wisdom, and a great deal of this found its way eventually to Northern Europe. Among them were alchemists and physicians like the Persian Rhazes (865–925) who wrote about measles and smallpox, and Avicenna (979–1037) who compiled a vast encyclopaedia which rivalled the works of Galen as a late Medieval textbook in the medical schools. Arabic science and medicine faded in the tenth and eleventh centuries for the same kind of reasons that science did not flourish in the Christian era. Islamic society could not support scientific activity in a declining economy, and its religious leaders became suspicious of the effect of such learning on the faithful, fearing it might lead to loss of belief in the Creator, and loss of direct dependence upon His will. The situation in Christendom, with the notable exception of thinkers like Abélard of Paris (1079–1142), was very similar, until the sixteenth century.

In 1537, at the medical school of the University of Padua, the tide turned for medical science, as it did for mechanics and astronomy there a few years later under Galileo. Andreas Vesalius (1514–64) the new, young professor of anatomy set in process a movement which was to explode the dogmas and fancies of the traditional science, and map out the application of modern scientific principles and methods in the realm of biology and anatomy. This work was to be greatly aided by the new techniques of printing, and the new styles of realism in art which made descriptive illustration effective as a means of communication alongside the printed word.

Vesalius' material was taken directly from the dissecting-room, and while he built upon the work of Galen, he greatly increased the range of knowledge about the human body, and the diagrammatic terms in which it was expressed. Vesalius showed some of Galen's errors—a dangerous thing to do to such an accepted authority—even if he made some mistakes himself. He detached the science of anatomy from the teleological (purpose-directed) arguments of his master, making it less clouded by 'theology'. Above all he looked at the human body itself, instead of inferring its nature from animal anatomy or traditional hunch; then he made precise wood-cuts of what he found. It was this line of investigation which was new, and which overthrew so many of Galen's ideas. Once again, as in the case of Copernicus' work, a basic new procedure of modern science had arrived.

Vesalius was not the only anatomist who practised human dissection. There is evidence that Michaelangelo and Raphael may have done so—certainly Leonardo da Vinci did—all illegally. In the sixteenth century there were several able anatomists, a flow of new-style texts and wood-cuts soon began to appear. The transition from pragmatic discovery made on the surgeon's bench to the systematic study of anatomy had begun.

The work of the anatomists went hand in hand with the growth of a new naturalism in art, which in turn was linked to realistic descriptions in botany and zoology. A glance at the development of the art of the early Italian Renaissance painters, in contrast with that of the Middle Ages proper, shows this naturalistic movement clearly. Botticelli's **Primavera**, painted in 1487, shows a strong interest in natural representation; it is full of real flowers and plants of different kinds. Leonardo's muscular men and animals show a rare observance of that which he sketches or paints. Dürer's exquisite animals show the influence of this mood over the Alps in Germany. With this movement came the art of botanical drawing through the German school: Otto Brunfels (1489–1534), Jerome Bock (1498–1554) and Leonard Fuchs (1501–66). Others observed, described and drew animals: William Turner (1510–68), who produced the first modern ornithology; Pierre Belon (1517–64) who produced books on fishes and birds; Guillaume Rondelet (1507–66), friend of Rabelais who described the fishes and marine life of the Mediterranean; Conrad Gessner (1516–65) the Swiss naturalist, who wrote a five volume **Historiae Animalium**; Thomas Moufet (1553–1604), a Londoner who did the first major work on insects.

It has been said that practical medicine has played as great a role in the development of biology as has technology in the evolution of the physical sciences. Certainly many of the founding fathers of botany, zoology, anatomy and physiology were medical men, and as we shall see, modern chemistry owes much to physicians too. This is partly because no university courses were available in the biological sciences outside the medical schools until the mid-eighteenth century. The desire to classify and describe with increasing accuracy stemmed from the ancient interest in herbs and healing. But the progress of the sixteenth century naturalists in mapping the structure and distribution of the creatures in the living world aroused new questions. Where does life originate? How is it reproduced? What determines the maintenance of the structure of a species? These were questions to be tackled, among others, by Francesco Redi in 1660 and John

Ray in 1693, and later by Gregor Mendel in his work on inheritance in 1857–68.

The achievements of the sixteenth century also stimulated investigation into other problems of anatomy and physiology for which no explanation was to be had. Of such a nature was the work of William Harvey (1578–1657) on the circulation of the blood. Having available a great deal of empirical, anatomical information, Harvey produced an explanation of how it fitted together in the circulatory system. He passed from anatomy to physiology by means of a theory he could test. Although the phenomenon of circulation was known to the Chinese, propounded by Galen and observed by Fernel, it was Harvey who provided the mechanical explanation of the heart's function, and who suggested that arterial blood carried some reviving factor while the venous blood returned without it. Once again an imaginative and testable theory founded on the evidence, ousted metaphysical conjecture based on ancient hunch or piety. Mechanism had arrived in biology as in the solar system. Religious explanations had to find another place in the scheme of things. The scientific method in this field also had come to stay.

2
From Alchemy to Chemistry

One of the great teachers of the early Church who exercised a profound influence on the Middle Ages—and the Reformation—was St. Augustine of Hippo (354—430). In his many writings, Augustine took up a distinctive attitude towards Greek thought, including Greek science. The danger of such study for him was that it was preoccupied with the creation and forgot the Creator:

> *Yet is it folly to doubt but he (the believer) is in a better state than one who can measure the heavens, and number the stars,*

and poise the elements, yet neglecteth Thee who has made all things in number, weight and measure.

The Confessions *V. 4.*

Here and elsewhere in his writing, Augustine was attacking purely contemplative knowledge—the idea of knowledge for its own sake. He condemned vain curiosity. Scientific knowledge, he held, was to be a way of worshipping God, and should be thought of as the handmaid of theology. This idea was generally accepted in the West during the Middle Ages by writers such as Hugh of St. Victor and Roger Bacon. The same ideas persisted strongly in Church circles through the Reformation, being reproduced by Luther and Calvin. They are held by many Christians today.

Scientific knowledge, such as it was, came to be thought of as a parable of spiritual or moral realities: a sacrament of divine grace to the faithful. No rigid distinction was drawn between scientific fact and spiritual truth or moral (divine) law. Augustine himself drew extensively on natural ideas in his theological teaching. He used analogies drawn from the human mind and from experience of light and illumination in his teaching about the Trinity. Similar analogical and allegorical use was made of other scientific ideas by his successors. This encouraged the widespread belief in certain types of ancient 'scientific' explanation such as the four-element theory of matter. The relations of the four elements—earth, air, fire and water—corresponded in some way to the relations of body, spirit, intellect and love. The facts of chemistry were expressed in these terms, and were used in this way by poets and moralists, even as late as John Donne (1571–1631). Such a unity of material and spiritual notions went deep and persisted for a long time. It provided the intellectual soil for alchemy to grow.

To modern minds, alchemy is a bizarre affair; but it was not always regarded as such. It emerged as a significant prescientific occupation during the Middle Ages, and represented a serious attempt to understand and exploit some of the mysteries of matter. Its overthrow by the gradual growth of atomism and ideas about the empirical properties of different elements provides a vivid illustration of the way in which the unity of spiritual and material was lost with the rise of modern science. Alchemy was the attempt, begun in c. 200 B.C. in Alexandria, to fit together theoretical interpretation and practical experience in a single, technical system. Its theoretical notions were drawn from the Greeks, particularly the Atomists. Leukippos (c. 500 B.C.) propounded a theory that all matter was made up of indivisible atoms or units in constant motion and separated by the 'void'. All atoms, said Leukippos, were made of the same stuff, but they were of different shapes and sizes. This theory was also taught by Democritus (460–370 B.C.) and reappears in the writing of Epicurus and, later, Leucretius. Democritus was apparently the first to distinguish between primary physical qualities and secondary or sense qualities such as taste and colour.

To these ideas, the early alchemists added Aristotle's Principle of Development—the notion that all material things naturally change and develop into an 'adult' form. Thus matter was treated as though it were organic rather than inert or static. From these ideas, and from their observation of geological changes taking place in the earth, the alchemists concluded that development occurred in nature: their art was to reproduce this in the laboratory. The apparatus used by the alchemists was therefore designed to create those conditions which would accelerate the natural 'developments' of minerals—the furnace, the retort, the 'nourishment' for base metals to be transformed to the higher or noble metals such as silver and gold. Their quest was to recreate the 'embryology of the earth' and achieve transmutation. In this quest they mastered many important techniques such as the controlling of furnaces, and they discovered innumerable chemical facts. To aid the process, they looked constantly for agents to accelerate this natural development—the catalysts—and above all for that crucial one, the Philosopher's Stone, which would turn all to gold.

In addition to this conceptual framework for their efforts, the alchemists adopted another element from Greek thought: the idea that everything had a soul and a body. Thus another aim of alchemical practice was to separate the souls and bodies of minerals, or to make mineral bodies 'incorporeal'. Naturally, gases or vapours were identified with the 'incorporeal' nature of substances, and the discovery of processes for the reduction of oxides led the alchemists to believe that they had found ways of both separating and recombining (in a purified form) the souls and bodies of created things. The religious significance of this was, of course, considerable. In Alexandria, where alchemy began, Greek ideas abounded. So too did the practical experience and craft of metal refiners, dyers, and merchants of the ancient Middle East. Added to these was the religion of Egypt and later, Islam and Babylon. Alchemy fused these ideas into its search for the mysteries of transmutation and perfection, and established itself as a useful spiritual activity in such a way that it survived into the fifteenth century. Unfortunately, in its later years, alchemy attracted to itself credulous fools and unscrupulous quacks of the kind lampooned in Ben Johnson's play, **The Alchemist.** In spite of this it would be a pity to ignore the attempt of the earlier alchemists to understand the processes of matter, and foolish indeed not to acknowledge the wealth of chemical data and technical skill which the unsuccessful quest of alchemy passed on.

The Church of the Middle Ages was highly suspicious of alchemy. The whole idea contradicted Augustine's teaching. The long association of alchemy with Islam made it of ill repute. The search for gold, for the 'spirits' of things, and for the device of the Philosopher's Stone smacked of witchcraft and evil. It is not surprising that alchemy was thought of as one of the Black Arts. Nevertheless, alchemy was practised, and by churchmen like Roger Bacon and Albert the Great, although they both found themselves at odds with ecclesiastical authority for it. In 1317, Pope John XXII published a bull **Spondent pariter** condemning the art, inferring that all alchemists were evil sorcerers. Further bulls in 1437 and 1445 from Pope Eugene IV exhorted the

Inquisition to be more effective in discovering and punishing 'magicians' and witches. The most severe appeared in 1484, proclaimed by Innocent III. The occasion was used by churchmen to begin a reign of terror in Germany to destroy magic and witchcraft. Even the attempt to label more genuine scientific study in alchemy and physics as 'white magic' did not preserve it from the worst periods of the Inquisition. This was one contributory factor in the delay of the scientific revolution in the field of chemistry.

The development of chemical theory originated not in alchemy directly but in Iatrochemistry—medical chemistry of the physicians of the fifteenth century. This group of men held that the function of the chemist was not to turn base metals into gold, but to find medicines to cure human ills. The terrible disasters of the years of plague over Europe in the fourteenth century gave impetus to this new direction of investigation.

The first to propound this idea was the Swiss physician Bombastus von Hohenheim (1493–1541), better known as Paracelsus. An outspoken and forceful character 'bombastic', von Hohenheim argued that the human body was basically a chemical system and needed chemical remedies for its ailments. The remedies he—and others—prescribed were wholly empirical ones—and often contained dangerous preparations such as salts of antimony, arsenic and mercury. Like the alchemists, the iatrochemists were inclined to mysticism, and believed in the influence of the stars. They looked, not for the Philosopher's Stone, but for the 'Elixir of Life' which would give health and long life.

For all its dangers, iatrochemistry, which lasted for nearly two centuries, advanced the sum of empirical knowledge, and led to the publication of **Pharmacopiae**—encyclopaedic records of medicines and their uses. Meanwhile further data, this time from Islamic and Byzantine sources, were becoming available in western Europe. The **Book of Fires** of Marcus Graecus (Mark the Greek) and the **Book on Colour-making** by Peter of St. Omer (c. 1300) are examples of the literature of the time containing chemical recipes handed down from the ancient past. In these ways, supplemented by the growing commercial interest in the refining of metals and manufacture of dyes and soap taking place in the late Middle Ages, the background was prepared for the emergence of chemistry.

This new science had to wait upon the adoption of a more rigorous experimental attitude than the alchemists had shown. In their writings, little attempt was made to distinguish between the false and the true. The efforts of the iatrochemists were often little better, but among them was one who did develop a new chemical theory, attacking the notions of both Aristotle and Galen. His name was Johann Baptista van Helmont (1577–1644).

Van Helmont thought that there were only two elements, Air and Water. All solid bodies, including the earths, were generated from water by the action of seeds or ferments, divinely created. This provided him with an explanation for his observation of a tree which gained 164 lb. in weight in five years without the lessening of the earth in which it was planted: all it appeared to need was water. He also introduced the idea of 'Gas'—a form of water existing in the extreme upper air, into which substances or vapours were converted by heat. This idea provided an explanation for several phenomena such as the explosion of gunpowder, the effervescence of fermenting grapes, and various other fumes. He then began to distinguish between different kinds of 'fumes' such as what we know as carbon dioxide, sulphur dioxide and nitric oxide.

Van Helmont's work was carried on by others, notably the philosophical chemist, Robert Boyle (1627–91). Boyle adopted the mechanical or corpuscular theories associated with the name of René Descartes. He was a first-rate experimenter in physics and chemistry and his books are full of descriptions of practical research. In spite of this, they show that Boyle's object was to formulate a new and adequate chemical theory to account for the large quantity of data which was by then available.

In his book **The Sceptical Chymist**, Boyle disposed of the four-element theory and demolished the principles of the 'orthodox' chemists of his day. His work is a good example of the negative criticism of existing theories, which Professor Karl Popper maintains is the only way in which any scientific theory can be refined and ultimately established. On the positive side, Boyle proposed that the basic form of matter was little particles which were organised into minute masses or clusters, some of which were indivisible. Particles of two sets of corpuscles might recombine to form a new indivisible compound under certain conditions, just as sand and ashes formed glass when heated strongly. Resistance to analysis by fire or chemicals was not an adequate criterion by which to say that a substance was elementary. This view was to be challenged later by Lavoisier, and Boyle's attachment to it prevented him from developing an adequate theory of primary elements. He concentrated his attention on the changes of corpuscular structure, and the reshuffling of particles in the formation of different compounds. His work on gases introduced him to other aspects of molecular and kinetic phenomena and the area of his work covered many aspects of modern physical chemistry.

Boyle frankly recognized that his corpuscular theory was an hypothesis to explain the data. It was an idea to be modified if new evidence came to light. The hypothesis enabled him to bring chemical theory into the new corpuscular philosophy of nature. These ideas were to prove particularly popular and fruitful in England in the seventeenth century, especially with regard to combustion and calcination. They were developed by men like Robert Hooke and John Mayo.

Although they were convinced of the existence of elements, the chemists of the eighteenth century found it difficult to relate the findings of laboratory experiments to such a theory. They had to account for changes brought about (apparently) by heat alone in combustion, or in the reduction of oxides to metal, and the affinities of substances for each other which seemed to cause chemical reactions to occur. The Phlogiston theory and the Principle of Affinity were the concepts introduced to make sense of these phenomena.

Phlogiston was the 'substance' emitted during combustion and calcination of metals. Some substances were thought to be very rich in it, such as charcoal, sulphur and phosphorus, others less so. When a metal was heated strongly, phlogiston was driven off, leaving a calx behind. The only problem here was that the loss of phlogiston in this case went with an **increase** in weight of the calx (by oxidation). It was even suggested by one of the chemists of the time (G. E. Stahl) that phlogiston, being a 'material fluid' had negative weight! It was compared with magnetism and gravity. Not until the chemistry of gases had progressed through the work of Black, Cavendish and Priestly, and the genius of Antoine Laurent Lavoisier (1743–94) had found a better explanation, was this concept eliminated from the chemical scene. Nevertheless, it was temporarily a very useful notion, and helped to establish chemistry as a distinctive science, and encourage enterprising men to turn it to profitable use in industry, especially in France and England.

Lavoisier was a skilful laboratory worker, though not in this field the most original. His gift lay in using the balance, and in basing adequate and simple explanations on what he found by repeating others' experiments. He was in correspondence with Black and Priestley, and on the basis of their work, devised a theory which had no need of phlogiston. He upheld his convictions that in natural and artificial chemical processes, nothing is lost and nothing is created (the principle of conservation), and that chemical substances conform to a natural order, as do living things. The first conviction led him to the idea of mass, and of quantitive experimentation. The second led him to ignore elements and atoms and concentrate instead on substances which had so far proved indivisible in the laboratory. With the aid of these definitions and methods, Lavoisier proceeded systematically to survey a wide range of chemical substances, analysing their constituent 'principles' (for example, oxy-gen, the acidifying principle) and comparing them. He introduced a new system of names—carbonic acid, black oxide of iron, sulphuret of arsenic, etc.—to indicate the fundamental make up of each substance.

He also distinguished between **physical changes** of substances in shapes, colours, or pressures of gases; **chemical reactions** involving the combination or separation of substances and the production of new ones; and **changes of state** between solid, liquid and gaseous conditions. Eventually he abandoned his early idea that heat could be combined (as caloric) with material bases even though it could not be weighed.

Lavoisier showed by his experiments of heating mercury that air contained oxygen not phlogiston, and the best explanation of the observed changes was the formula:

Mercury + Oxygen (respirable air) = Red calx of mercury (mercuric oxide)

Lavoisier's thesis was extensively tested using other metals and oxides. Many accepted his explanation; others, like Joseph Priestley, did not. Further experiment established it, however, together with the quantitative methods which Lavoisier employed. In this way, the idea of mass (or quantity) was introduced into chemistry, which up to Lavoisier's time had been dominated by qualitative analysis. Lavoisier also put the theory of affinity between elements on a proper, scientific basis, opening up new areas for subsequent exploration in inorganic and organic chemistry.

It was John Dalton (1766–1844) who revived the atomic theory and combined it with Lavoisier's idea of quantitative affinity between elements. He drew up the first atomic table of twenty 'atomic weights'. As a result of this sequence of development, chemistry became separated from physics as a distinct science until, through the work of Berthollet, Avogadro, Prout and Mendeleev, a deeper synthesis of concepts at the level of atomic theory emerged.

As with physics and, to all intents and purposes, astronomy also, the rise of chemistry produced a self-sufficient pattern of explanation for an area of phenomena. The unity of alchemy, iatrochemistry and the Medieval world-view was broken; a new language with distinctive concepts and theory arose for this area of experience; a new and powerful tool was added to the industrialists' equipment; and the direct involvement of religion in all things was seemingly pushed back, or as some came to think, eliminated altogether. It fell to the philosophers to make an attempt to reshape the total map of understanding as best they could, and to a small number of theologians to work out the real implications of the revolution which had occurred.

Meanwhile, Western Christendom in its official Catholic and Protestant attitudes carried on as though little had changed. The science of the bible and the ancients was good enough. The terminology of the Medieval synthesis was still accepted on either side of the Reformation fence. The attitude of St. Augustine served the clergy and ministers well enough, even if merchants and craftsmen in their flocks thought differently about the new sciences. Thus the split mind of modern Christendom was born, and the foundations laid for the clash of authority which came in the nineteenth century, and which is with us still.

3
The Fracture of Knowledge

The world of the late Medieval West shared a common understanding. It accepted, within limits, a common cosmology—that is a model or picture of the universe. Ptolemy's pattern of the planets was interwoven with a three-decker spiritual order of heaven, earth and hades. Material things were thought to be organized by spiritual principles, and thus provided analogies of them. Under the influence of Aquinas' thought, due place was given to reason, knowledge provided through the senses (cognitive knowledge), and mystical or intuitive knowledge. Each had its sphere, and all contributed to the unity of knowledge of the essentially unitary system. Penetrating all things was the providence of God, the source and end of knowledge, wisdom and truth, whose purpose was disclosed in the least of creatures as in the sum of all. Knowledge of him might be discovered in part by reason, but only in full from Holy Church, whose institution and teaching expressed the mind of Christ himself the Son of God.

The rise of the sciences broke this rounded pattern. As we have seen, in cosmology, the material explanations of mechanics displaced the purposive explanations of theology. The realistic descriptions of the anatomists did the same for the human body, even causing speculation on the place within the body of the soul! The use of mathematical models drawn from algebra, arithmetic and geometry replaced the ancient categories of motion and divine law for the fields of optics, projectiles and the movements of the planets. In chemistry and physiology new concepts were introduced in place of the old ideas of humours, earths, spirits and transformations; and these new concepts were related to the new mechanics of everyday particles rather than to the religious convictions of the Church, or to the simple psychology of the ancients.

At the same time, theology and moral philosophy struck out for new forms. On the one hand the Reformation revived the doctrines of St. Augustine and gave added authority to the newly translated terms of the scriptures, creating a reformed biblical theology much removed from scientific thought. On the other hand, rational theology—with a reasonable interpretation of religious ideas—arose in England with the writings of men like Lord Herbert of Cherbury and the Cambridge Platonists.

Meanwhile, the Cartesians had separated mind from mechanical nature, and defined an exclusive area for philosophy and religion. The province of ethics was mapped out by Spinoza, who listed new moral goals in place of those laid down by the divine law within the Medieval scheme. Mind and morality were thus separated from intimate connection with the natural order and the objective knowledge of given spiritual or moral realities within it. This laid the path to the rejection of 'scientific' or cognitive ways of knowing such things on the basis of commonly accepted evidence.

The great philosopher who trod this path to its logical conclusion was Immanuel Kant (1724—1804). Kant rejected metaphysics—total systems of explanation for the whole of knowledge and experience—though he affirmed the existence of a moral order in the world. God's existence could not be proved by reasoning from definitions of what the word 'God' means, or from the evidence of the world around. Only the inner compulsion to make a moral choice or the emotional response to the mysteries of the universe, he indicated, could justify belief in God. The categories of science, such as space and time, could not be applied outside the realm of things that were experienced through the senses. Thus Kant carried to the limit the fragmentation of knowledge into separate compartments.

Meanwhile the world of science itself was sub-dividing into distinct sections: mechanics, astronomy, anatomy, physiology, biology, chemistry. By the end of the nineteenth century, when the classical system of Newtonian physics had reached its fullest extent, even the sciences seemed like the fragments of a stained glass window. No one could be sure of the pattern except (as they claimed) the triumphant men of physics.

By then, such was the confidence of different scientists in their 'objective' explanations, that men could forget those other kinds of knowledge or deny that they were real in any sense at all. Scientific man forgot the culture which had brought modern science to birth, and dismissed those other elements of knowledge and imagination in the history of the scientific enterprise as worthless. Seeking a complete and final system, the brightest fragments were taken to be the whole picture. In this way scientific man became something of a slave to the creations of his mind, even though he did not know it. Thus the questions posed by this story of the fracture of knowledge are: can we ever isolate one kind of knowledge from another? Is not the science (or theology, or philosophy, or moral system) of every age a reflection of its culture and social attitude? Must we not seek to bring together artist, scientist, philosopher, theologian to gain an understanding of knowledge—and of culture—in our own time?

4
Topics, Questions and Sources

1
What can you discover about the aims and methods of physicians and surgeons in the period c. A.D. 1200—1600?

2
Make an exhibition indicating the interests and achievements of any one notable pioneer in the field of science during the period covered by this section.

For discussion:

3
Was the Church right to forbid dissection in the fifteenth and early sixteenth century?

4
Was alchemy nonsense?

5
Do you agree with St. Augustine that science should be the handmaid of theology?

6
Debate the motion: 'In the opinion of this house science is a matter of fact—all else is supposition'.

Sources:

A History of Biology C. Singer *Schuman* 1952.
A Short History of Medicine C. Singer *O.U.P.* 1962.
Origins and Growth of Biology (ed.) A. Rook *Penguin* 1964.
A Short History of Chemistry J. R. Partington *Macmillan* 1937.
Alchemy E. J. Holmyard *Penguin* 1968.
The Architecture of Matter S. Toulmin & J. Goodfield *Penguin* 1965.
A History of Western Philosophy Bertrand Russell *Allen & Unwin* 1961.

C
Key Factors in the Scientific Revolution

1
The Importance of Technology

The majority of methods available to scientists in the first stages of the scientific era were simple, unsophisticated and derived very largely from the various crafts of the time. Until the sixteenth century, the facts of science were accumulated not only by those who might be called the men of science, but by craftsmen, physicians, surgeons, travellers and navigators. This helps to explain why so many of the key figures in the rise of modern science before this time were theorists using known data rather than 'laboratory' men.

Galileo was among the first to attempt new and refined ways of obtaining scientific data, in his use of the telescope for astronomy, by his study of materials, and his devices for the measurement of time. He was prominent also in his use of the techniques of mathematics, as we have seen.

The development of scientific instruments proceeded through the cooperation of scientist and craftsmen, often in the interests of commerce—as in the case of the improvement of clocks and compasses by men like Robert Hooke and the clock-maker Thomas Tompion. Robert Boyle was emphatic about the need for students of natural history to learn from trades and crafts. From the use of the balance in assaying came its use in quantitive chemistry. From the skill of artisans in the commercial field of chemistry, metallurgy and mining came both problems to be solved and also techniques of refining, heating and synthesis for chemists and physicists. Technical books, such as the **Treatise on Ores and Assaying** (1574), the **Distillation Book** of H. Brunschwig (1512), and **De re Metallica** of Agricola (1556), made details of techniques available to the scientific world. During the seventeenth century the whole concept of science changed, and the invention of specific scientific instruments began. These included the chemical balance, the astronomical telescope, the microscope and the air-pump—all of which used principles already known by the craftsmen. A rapid improvement in glass technology made possible the grinding of better lenses and the manufacture of better glass vessels. Scientists like the great microscopist Leewenhoek became experts in instrument manufacture, producing precision lenses and high resolution microscopes for the laboratory.

During this stage, a new type of instrument appeared: this was the measuring instrument. The vernier scale was attached to telescopes; so were devices for measuring angles with greater accuracy. Micrometers were brought into use. Toricelli's barometer, invented in 1643, was adapted in 1660 for the measurement of atmospheric pressure (giving birth to the science of meteorology). Galileo's thermoscope, consisting of a glass tube ending in a bulb at one end and a bowl of liquid at the other—which he used for demonstrating the expansion of air when heated—developed into the thermometer. From the first calibrated type in 1665, this was further improved by Farenheit's scale in 1715 and the centigrade scale in 1743. In this way instruments first used qualitatively—to demonstrate certain kinds of phenomena—were adapted to quantitative use.

In the case of later sciences, such as the study of electricity, no practical or theoretical development could occur until the necessary instruments had been prepared to provide basic data.

The same story applies to many recent sub-divisions of scientific study: biochemistry, biophysics, the science of materials, crystallography, and so on. The influence of technology was therefore direct and constant in this way, and has been so ever since. But in another way, technology had a more subtle effect.

In the sixteenth century a major shift of interest occurred from the metaphysical system of the medievals to the narrower field of physics and particularly mechanics. This, as we have noticed, was reflected in the literature and art of the period as well as in other aspects of its culture. The metaphysical poets such as John Donne and Henry Vaughan were succeeded by men like Milton, Dryden and Pope, whose writings laid a new stress on human reason. Alexander Pope, in his **Essay on Man** could rhyme:

Know then thyself, presume not God to span,
The proper study of mankind is Man.

Coupled with this interest in reason was a growing concern for business, manufacture and machines. Intelligent men were beginning to realize the commercial possibilities of knowledge. In the background of this major shift of cultural emphasis was the awareness, expressed by Francis Bacon, that 'Knowledge is power'. This awareness was brought about by the advances in technology already made in western Europe. The Age of Reason coincided with a major economic upsurge in England, France and other European countries, and with a major expansion of trade abroad. The influence of technology on the whole intellectual climate is thus a key factor in the rise of modern science.

2
The Growth of the Scientific Community

Many of the new steps to bring about the rise of modern science were taken by individuals, working largely on their own. There are important exceptions to this—notably in the case of the two giants, Galileo and Newton. Galileo worked, first in Pisa and later in Padua, in the context of a growing concern about the anomalies of Aristotelian science, and with the aid of colleagues, many of whom no doubt disagreed with him, who shared in the development of the new mathematics and who constituted a community of enquirers. Newton, although an individualist in many ways, worked upon the written documents of others, and enjoyed discussion with pupils and colleagues. The same is true of Vesalius and of William Harvey. Nevertheless before the middle of the seventeenth century, groups of learned men were rare, and the publication of scientific journals had only just begun. In many cases, the universities were institutions for the preservation of learning in the old tradition rather than centres of research and investigation: the new sciences prospered outside them.

The Royal Society in England (1662) can trace its antecedents back some way. Like the *Academie des Sciences* in France (1663) it took its origin, as did other less-known groups such as the members of Leyden botanic garden (c. 1580) the *Jardin de Roi* in Paris (1579) and the *Accademia dei Lincei* in Rome (1603), from the interest of educated and wealthy persons in the new sciences. This interest was stimulated by royal patronage, by discoveries in the newly explored territories in East and West in the sixteenth and seventeenth centuries, and by better arrangements for travel and printing. Among the members of such societies were landed gentry, doctors, clergymen, lawyers and soldiers.

These societies were as much concerned with philosophy as with practical experimentation. Important as experiment was—and increasingly became—in the rise of the scientific movement, it had to be related to theory, to new concepts and to general philosophical ideas for any leap forward to occur. This is particularly clear in the history of the Royal Society. The foundation of this institution under the royal charter of Charles II in 1662 owed much to the writings of the philosopher Francis Bacon (1561–1625) in his books, **Novum Organum**, **The Advance-**

ment of Learning and **The New Atlantis**. Thomas Sprat, the Society's first historian, said of Bacon that he was the

> *One great Man, who had the true Imagination of the whole extent of this enterprise, as it is now set on foot.*
>
> **History of the Royal Society** p. *35*.

The religious ideals of Bacon—his convictions about God's rational creation, his linking of truth with utility, his stress on humility before facts, the combination of empirical evidence with a general theory in his idea of human knowledge—were all adopted by the Society. Under the influence of Bacon's version of the natural philosophy its early membership transcended the denominational barriers of the Reformation. As a result, the Royal Society at its foundation was the first interdenominational Christian organization, having Anglican divines, presbyterians and papists among its Fellows.

Alongside societies for the discussion of new ideas arose a different type of organization such as the *Accademia del Cimento* in Florence (1657–67) which—as Francis Bacon had suggested in the **New Atlantis**—was a kind of research institute with members gathering information and working collectively on scientific problems. Among the members of this academy was Torricelli, the discoverer of barometric vacuum. These academies were able to gain the services of the best instrument-makers of the time, they carried on scientific education of young men, and encouraged the publication of experiments and new ideas.

These societies and academies also produced journals: the **Philosophical Transactions** of the Royal Society, started by its first secretary, Henry Oldenburg, and the **Journal des Scavans** in France are early examples of such efforts. These publications in turn encouraged men in other countries to work for the establishment of scientific organizations. One of these was G. W. Leibnitz (1646–1716) who struggled for forty-seven years before his dream of the Berlin Academy of Sciences was established.

As well as academies concerned with physics and chemistry,

this period is remarkable for the establishment of the botanical and zoological garden as a royal or civic institution. Such places were usually the centre of scientific groups and a good deal of observation and experiment on various scientific topics. Observatories flourished under royal patronage—supporting the science of astronomy. The investment of Charles II in John Flamsteed's observatory at Greenwich proved a sound one: English navigators were to enjoy the fruits of its work for centuries to come.

One marked effect of the societies and academies of science of the seventeenth century was the rapid diffusion among the scientific community of the new 'natural philosophy' in Europe. Critical thinking and mechanistic ideas spread into every corner. They are reflected in much—though not all—of the thought of Hooke, Boyle and Newton, tempered by the growing requirement that such notions should satisfy the test of experiment. Newton, unlike Descartes, subscribed to the notion of the aether and, of course, to gravitation. After his time, the Royal Society moved away from many of Descartes' tenets in its thinking. Nevertheless the influence of the organizations of science extended to the spread of philosophical and religious notions, as well as to the promulgation of the facts and hypotheses of science.

3
Key Religious Factors in the Scientific Revolution

Many factors contributed to the growth of the scientific method and the rise of the modern sciences. The period between the twelfth and seventeenth centuries saw a remarkable conjunction of events with an equally remarkable collection of minds. It was a period of exploration in every sense; geographically, intellectually, socially, and in terms of human nature and dignity. Of critical importance was the growth of technology and the shift of cultural life from the knights and monks of the early Middle Ages to the bourgeoisie of the towns. Capitalism encouraged individual enterprise and effort which was rewarded commercially and by the royal court for reasons of prestige. The hard reasoning of the merchant class fostered rationality and criticism —the seeds of science—outside the conformity of the feudal or ecclesiastical order. But yet there was something more. All these factors had operated in other civilizations before: in China or Greece or Islam. Yet in none of these civilizations did the scientific explosion occur. This fact has forced scholars to consider the possible influence of religion on the rise of modern science, and many—but not all—consider it to have been a crucial factor.

During the period which has been under review, it is evident that an ambiguous attitude was adopted to the rise of scientific thought by the Christian Church. For long, the Church held to the teaching of St. Augustine of Hippo, that knowledge should not be pursued for its own sake, but for the glory of God, and as the servant of theology. It was only with the schools of Charlemagne and the growth of the monasteries that western Christendom began to come to terms with large parts of classical learning and the attitudes they contained. In the eastern part of Christendom, the antagonism of the early Fathers of the Church to Greek thought was less marked than in the west, and in the tradition of Orthodoxy, a successful effort was made to penetrate what had been handed down from ancient Greece, to integrate such learning with Christian teaching in the schools, and to relate it to the Divine Liturgy and the Wisdom of God. This approach did not commend itself to the western Church for theological and political reasons, and the Byzantine tradition was first ignored and then, under Turkish pressure, dissipated.

In the period of our concern, however, some of this tradition reached the west. More important, scholars in the west had to come to terms with the Greek tradition because of the intellectual advance of Islam. In the thirteenth century, Islamic universities in Spain were setting the pace in ideas and religious apologetic for the faith of the Prophet Mohammed with the rediscovered ideas of Aristotle: the threat to Christendom was considerable. The foundation of the universities of Christendom and the establishment of the Dominican Order were part of the Church's response to this challenge. It bore fruit in the work of St. Thomas Aquinas, who stole the thunder of Islamic scholarship by 'baptizing' Aristotle.

This process recovered the intellectual integrity of Christendom on the one hand, and made explicit—and limited—the place of reason on the other. Aquinas' clear distinction between reason and revelation, his uniting of reason and devotion in a redefinition of faith, and his emphasis upon the rationality of God made the beginnings of the scientific movement possible as they could not have been before. Above all, Aquinas reinstated the doctrine of creation in Christian understanding. His reconciliation of the Prime Mover of Aristotle with the God and Father of Jesus Christ renewed Christian interest in the natural order.

The doctrine of creation, placed anew alongside that of redemption, implied that nature was rational, consistent and orderly. This became one of the unquestioned assumptions of modern science. It also suggested that the details of the creation could only be known by observation—something which Aristotle, but by no means all his classical fellow-thinkers, had recognized. Again, the doctrine of creation—as the Franciscans showed—implied a positive attitude towards God's world. In contrast to the otherworldly emphasis of the Dark Ages when life was particularly nasty, brutish and short, the new climate of faith recognized that all God had made was good, except the sin of man.

It is true that self-denying piety remained dominant and there was much preoccupation with death, judgement, heaven and hell. The paintings of Michelangelo and Grünewald show this. But Man's role in creation, spelt out in the early chapters of the Book of Genesis, was now taken as seriously as the myth of the fall. Further, with the renewed interest in the scriptures which came with the recovery of learning in the Church, there came also a new appreciation of the dynamic will of God. St. Thomas captured this in his teaching about God as the First Cause. His emphasis on the Divine purpose in Creation stirred men to action out of the fatalistic mood of earlier Christian generations. Under such influence the stage was set for the new movement to arise.

The development of nationalism, the middle classes and the capitalist system which aided the growth of the sciences has been traced by many historians—such as R. H. Tawney—to the influence of the Reformation. This was certainly a significant factor in encouraging independent thought in contrast with belief in the official world-view required of the Catholic world. It also fostered national interests and thus encouraged industrial enterprise and scientific research sponsored for reasons of prestige.

It is a fact that seven out of the ten founder members of the Royal Society were Puritans. Many of the most active English scientists in the late sixteenth and early seventeenth centuries were clergymen. The readiness of Reformation churchmen to take an interest in public and philosophical questions in the seventeenth century enabled them to absorb the new natural philosophy and the rational schemes of Descartes and Newton. This ecclesiastical encouragement to science and technology provided the cultural 'soil' in which the new understanding could grow.

Among the Calvinist churches, the so-called **Protestant Ethic** encouraged work as a good thing for man. It glorified secular vocations rather than religious ones, and emphasized the importance of the welfare of the community. The Christian was encouraged to glorify God by working honestly, saving hard and giving generously of his wealth, time and talents. John Wesley later crystallized this attitude in his slogan 'Get all you can, save all you can, give all you can'. Here are the seeds of capitalist economics and rugged private enterprise as well as of the welfare society. Scientific endeavour exhibited these protestant virtues, and was considered therefore to be divinely blessed. It would reveal God's handiwork and promote the good of his people. Man could thus understand God through nature, since nature was part of God's revelation to man.

In the religious circles of the Enlightenment, human reason

was regarded as God's greatest gift to man, enabling him to 'think God's thoughts after Him' and share his godliness. Joseph Addison's hymn expressed exactly the influence of Christianity on the culture in which modern science grew:

The spacious firmament on high,
With all the blue ethereal sky
And spangled heavens shining frame,
Their great Original proclaim.

The unwearied sun from day to day,
Does his Creator's power display,
And publishes to every land,
The work of an Almighty hand.

The Puritan divine, John Cotton, wrote in 1654 that the study of Nature was a positive Christian duty. In his last will and testament, Robert Boyle wished the Fellows of the Royal Society

> *a most happy success in their laudable attempts to discover the true nature of the works of God, and praying that they and all other searchers into physical truths may cordially refer their attainments to the glory of the Author of Nature and the benefit of mankind.*

Although it is true that on occasions the Church opposed the scientific revolution—sometimes with severity—it must also be acknowledged that Christianity contributed significantly to the attitudes and assumptions necessary for that revolution to occur. Let us not forget that almost all the revolutionaries were Christian men.

4
Topics, Questions and Sources

1
Prepare an exhibition to illustrate the influence of one technical development on the advancement of science.

2
The initials F.R.S. are highly prized: discover all you can about the formation of the Royal Society, and its past and present work.

Discuss:

3
'Modern progress in technology has far outstripped man's ability to control himself.'

4
'The scientific revolution has finally separated man from God.'

5
Read and discuss John Osborne's play **Luther.**

Sources:

The Scientific Revolution A. R. Hall *Longmans* 1962 *Ch. 8*
The Origins of the Scientific Revolution (Documents) Ed. H. Kearney, *Longmans* 1964.
Issues in Science and Religion I. G. Barbour *S.C.M.* 1966 *Chs. 2 and 3.*
The Rise of the Technocrats W. H. G. Armytage *Routledge* 1967 *Ch. 1.*

Part Two
SCIENCE AND RELIGION – THE MODERN DEBATE

A
Areas of Conflict

1
The Case of Charles Darwin

A fossil anteater seems an unlikely starting point for one of the most explosive scientific theories ever to shake the world, but according to his autobiography, it was Darwin's re-examination of this specimen—discovered on the voyage of the **Beagle**—which started the chain of events leading to his famous theory of natural selection. It was in Cambridge, in 1837, that Darwin noticed the similarities between this large, horse-sized fossil and small modern anteaters alive today. If in fact the fossil was the ancestor of the living animal, the accepted notion of the time that every species had remained just as it was from the days of Creation could not possibly be true.

As he continued work on his journal of the **Beagle's** voyage, Darwin discovered resemblances between living species which pointed in the same direction. Similarities between the finches he had extensively observed on the Galapagos Islands seemed to indicate a common origin of the species. Darwin was puzzled by the fact that some features—especially in their beaks—had been preserved and repeated, but others apparently ignored, in a different environment. The same problem reappeared as he studied his collection of South American fossils. Could there have been a process of change and selection in different conditions? Selection was to become his key word, but the establishment of his hunch required a vast comparative study of the anatomy of many different species.

It was a daunting task, hardly encouraged by the prevailing religious conviction that the Almighty had made the species according to a preconceived design at 9. a.m. on October 23rd 4004 B.C.—the date and time 'discovered' from the Old Testament by John Lightfoot (1602—75) during Newton's lifetime. This, together with Darwin's early inclination towards the Christian ministry, no doubt contributed to the slowness with which he set about compiling his theory and publishing it. The voyage of the **Beagle** took place from 1831 to 1836: **The Origin of Species** was not published until twenty-two years later.

Darwin's original hunch had set his mind working towards a solution of the problem presented by his observations. In 1838 he read Thomas Malthus' book **An Essay on the Principle of Population**, written in 1798. In this essay Malthus suggested that there was a constant struggle for survival in the human population against war, famine and disease. But for this struggle, he thought that the human race would overrun the earth. This idea, together with his familiarity with Charles Lyell's view that the earth was extremely old and had allowed for long, slow organic changes in fossilized creatures, provided the theoretical framework for Darwin's theory.

> *It at once struck me*, wrote Darwin, *that under these circumstances favourable variations would tend to be preserved and unfavourable ones to be destroyed.*

The age-long struggle for survival in any environment provided for the process of selection. The better adapted survived; the poorly adapted did not. In the long term, this process would bring about new species.

In 1842, Darwin moved with his wife to a country house at Downe in Kent. There he brooded, wrote preliminary essays for

The Origin of Species, and nursed his anxieties and illnesses. In 1858, he had drafted eleven chapters of a book, encouraged by his friends Sir Charles Lyell and Sir Joseph Dalton Hooker, then Director of the Botanical Gardens at Kew. But in June of that year, he received through the post a copy of a paper written by a young naturalist, Alfred Russell Wallace. Wallace had been working in Malaya and the East Indies; he too had put together the essential points of a theory of natural selection, and was seeking the opinion of a fellow explorer about its suitability for publication. Darwin was shaken, and wondered whether he should publish his own views in the light of this bombshell. Fortunately, both Lyell and Hooker reacted quickly and arranged for Darwin's paper to be presented to the Linnean Society at the same time as Wallace's. The effect on the learned society was stunning: both papers were received in grim silence. Nothing like this had occurred since the publication of an anonymous work in 1844 entitled **The Vestiges of Creation**—a speculative treatise which had suggested an evolutionary development of the human species, though without any empirical evidence to support it.

Darwin was now persuaded to prepare a book from the material he had already written. His publisher cut down the title to **The Origin of Species** and it appeared in 1859. The first edition sold out inside twelve hours. From this point the storm of public controversy broke about Darwin's ears as it had done over Lyell's theory some years before. **The Quarterly Review** accused him of contradicting the revealed relationship between the creation and its Creator! Others used the weapon of invective and insult to try to discredit Darwin. The **Athenaeum** magazine charged him with making the suggestion that man had descended from the monkeys into a kind of creed—though Darwin had by then said very little about the place of man in his scheme of natural selection.

The row continued during the following year and reached its climax when the British Association for the Advancement of Science met in June 1860 at Oxford. Darwin's theory was attacked by three speakers at the conference, and the Bishop of

Oxford, Samuel Wilberforce, came to address a packed meeting of 700 people on the question. At the end of a half-hour denunciation of Darwin's views in the name of scripture and faith, Wilberforce turned to Thomas Huxley, a prominent biologist, who was a known supporter of the theory, and asked bitingly, 'Was it through his grandfather or grandmother that he claimed descent from an ape?'

Huxley whispered to his neighbour 'The Lord hath delivered him into my hands!' and made his reply. At the heart of his answer, Huxley told the audience that he would feel less shame in having an ape for his ancestor than a brilliant man who plunged into scientific questions of which he knew nothing. Uproar broke out: the issue was joined.

Darwin himself was not present at this meeting, and thoroughly disapproved of the controversy his work seemed to arouse. In the book he had been at pains to avoid theological dispute, and used the words 'God' and 'Creator' in the final chapter, attempting to reconcile his theory with religion:

> *There is a grandeur in this view of life, with its several powers, having been originally breathed by the Creator into a few forms or into one . . . .*

Although he was later to call himself an agnostic, his orthodox beliefs were shaken more by the current crude theories of the atonement and of hell than by his own work. Even at the end of his life he remarked to a friend that the wonderful contrivances in Nature appeared to him as the expression and effect of mind. 'That often comes over me with overwhelming force,' he said, 'but at other times it seems to go away.' To the end he maintained, rather naively, that his own religious opinions were of no consequence to anyone but himself.

The Evangelicals, led by Bishop Wilberforce, had begun the attack on Evolutionary theory as contrary to the words of scripture, divine revelation and religious authority. It was sustained and developed by Tractarians like Dr. Pusey. Yet some churchmen of more liberal mind like Charles Kingsley, R. W.

Church, Aubrey Moore and F. D. Maurice, were able to distinguish between scientific theory and theological assertion and accept Darwin's conclusions.

Pusey's onslaught, for all its conservatism, was a subtle and important affair. He claimed that evolution had been taught long ago by St. Augustine of Hippo and was in itself no new thing. He admitted that Darwin had left room for a Creator who breathed life into a few forms, but he accused him of forgetting God after this initial act. Unlike Newton, Darwin had left no place for God to continue to act in the evolutionary process; all was now left to chance. This was the cardinal heresy of Darwin's thought, Pusey maintained, brought about by the error of overspecializing in science.

Pusey thus brought to light the perennial problem facing the scientist—that in the nature of his work he is bound to find natural explanations and eliminate the supernatural. If he does not do so his work is unscientific—as Darwin's references to the Creator were strictly unscientific statements. But if he does eliminate such references, he is constantly tempted to assume that his partial explanations and insights constitute the only kind of understanding of reality. Those who have recently attempted to overcome this problem, as Charles Raven and Teilhard de Chardin have, invariably fall foul of others in both scientific and theological camps. Yet the urge to make some complete all-embracing picture or metaphysic is there: without it, scientific man seems fated to live with a split mind, unaware of the subtle influences of one pattern of understanding on another, or of the fact that both science and theology are human activities of the same creaturely mind.

Darwin's view of man's evolution, contained in his second work **The Descent of Man, and Selection in Relation to Sex** (1871), appeared to contain a straight challenge to the Christian doctrine of the uniqueness of man. The idea that *homo sapiens* had arisen from lower forms of life seemed to his opponents to contradict the belief that man was a child of God, fallen from perfection, and intrinsically different from the animals. This time it was

Christian doctrine as well as scriptural authority which was in jeopardy. It took many years for theologians to spell out clearly that an ancient mythology contains a different kind of truth from that of a scientific theory; that man is unique even though he has evolved by a long process of natural selection; and that the authority of the bible or the Church could not reasonably be extended to govern the evidence or reasonable interpretation of scientific investigation. It is taking even longer to persuade some members of the scientific community that there are other kinds of authority in human life than those of the laboratory.

The refutation of the history of the first days of creation (in the two separate accounts in Genesis) by Darwin's theory has remained a sore point since his day for those who accepted the bible as an universal authority for every kind of truth. There are still those who believe, as Dr. Michael Ramsey has put it, that the bible descended from the skies, in English, some editions with the Apocrypha and some without! It is a continuing problem for theologian and ordinary churchgoer to distinguish in detail the theological truths of the bible from the limited human insight, understanding, language and cultural forms in which that truth was recorded and had been transmitted. The temptations to religious people to regard the cultural patterns of their basic documents as God-given and unalterable are very strong. The theory of natural selection played an important part in forcing Christians to abandon some of their cruder views, and in encouraging a more critical approach to the nature and meaning of the biblical literature.

Darwin's theory also challenged the widely held version of the argument for God's existence based on the notion of design in the natural order. From the time of John Ray's **Wisdom of God in the Works of Creation** (1691), largely reproduced by Archdeacon William Paley in his **Evidences of Christianity** (1794), this argument had rested on the detailed examination of specific organs and organisms, and their remarkable 'fitness' for the environment in which they lived. Since Darwin had now shown an interaction between organism and environment, making this fitness a matter of natural selection, this could no longer carry the weight which had attached to it. Before long, philosophers and theologians revived the argument from design in a different form, relating it to the whole notion of evolution—and landing themselves with the problem of waste, catastrophe and 'evil' in the process. The most important of these newer arguments have come from the realist school of philosophers including men like William Temple and A. N. Whitehead.

Whitehead's 'process philosophy' is undergoing a revival at the moment especially at the hands of American theologians. His scheme of thought (or metaphysic) stresses the primacy of time, the interrelationship of different events, and the idea of reality (including God) as an organic process. It is an important offshoot of the prevailing evolutionary thought of the late nineteenth century which was expressed in ethics, poetry, prose and even theology of the time (such as Newman's Theory of Development of Doctrine) of which Darwin's achievement was a part.

Charles Darwin, kindly, patient, Victorian to the core and detached from controversy in his Kent retreat, raised three huge questions for the debate between science and religion which has continued ever since:

What is the nature of scientific truth and of religious truth?

What is the proper relationship between them?

What kind of evidence can there be for the existence of a good, purposeful God?

2
The Problem of Life

The natural philosophy of René Descartes placed animals and plants, and even the human body, in the mechanical world. Only thought did not obey the laws of mechanics. No essential difference was to be found between the realms of animate and inanimate things, except in the mind of man.

Some were prepared to accept this far-reaching generalization, but others—the vitalists—felt that living things had something which non-living lacked. In biology an entity called the vital spirit or the life-force was proposed to account for those phenomena which distinguish living matter: reproduction, growth, response to stimuli, and self-regulation or maintenance. Vitalism was one of the first protests against the mechanical materialism of the seventeenth century. Biologists who subscribed to it were asserting their convictions, based on what they could observe, that whole organisms could not adequately be explained by the prevailing concepts of atoms and machines. Like the German scientist and poet Goethe (1749–1832), who recognized that the whole organism was more than the sum of its parts, they believed that it could not be understood or described by dissection alone. Following Aristotle, vitalists clung to their own language: that of purpose, response, and adaptation.

Georg Ernst Stahl (1660–1734), father of the phlogiston theory in chemistry was the parent also of post-Aristotelian vitalism. The laws governing the living body are, he said, not the laws of mechanics but the laws of the 'sensitive soul'. He believed that such laws operated through chemical processes in the body—an anticipation of later chemical physiology.

John Hunter (1728–93) and Joseph Priestley (1733–1804) were notable vitalist experimenters of this period, and down to the present day, a number of biologists subscribe to a modified view of vitalism. They maintain that there is an irreducible reality belonging to whole organisms or plants which cannot be found in inanimate forms.

The difference between a modern vitalist and a seventeenth-century vitalist is that, nowadays, nobody seriously supposes that there is a mysterious 'thing' in a living organism which does not obey chemical or physical laws. Such a notion has been eliminated by the emergence of the life-sciences—biochemistry, bio-physics, plant and animal physiology. Practically all would subscribe to the view that these sciences, developed from the time of Justus von Liebig (1802–73) and Claude Bernard (1813–78), have shown that living organisms are distinguished from non-living by having more developed levels of structure and bio-chemical organization.

The difference nowadays between a vitalist and a non-vitalist is focused on the use of purposive language—language indicating directiveness of organic activity. For example it is possible for a vitalist to speak of eels trying to swim to their spawning grounds in the Sargasso sea: a non-vitalist would prefer to describe the pattern of behaviour of the eel as a response to certain stimuli associated with the warmer water of the Gulf stream. In point of fact, the scientific evidence is identical but the interpretations use different kinds of language: the vitalist stressing the special status of the living organism and implying some kind of deliberate effort: the non-vitalist using a mechanistic or chemical explanation without any suggestion of the creature being conscious of its action.

The strength of the physico-chemical pattern of explanation is that it is usually capable of scientific verification, and lends itself to the development of wider explanations for animal or plant behaviour. The vitalist explanation is descriptive in readily understood terms, but it does not provide a model which can be verified or developed. For this reason the physico-chemical type of explanation is rapidly capturing the field, though with increasing regard to the observed or inferred complexity of organization of living things.

This process has leapt forward recently with the isolation of viruses which exhibit some, though not all, of the characteristics of living organisms. They will reproduce in host cells, grow, respond to certain stimuli, maintain and adapt themselves, producing strains resistant to antibiotics for example. At the same time some, like the tobacco mosaic virus, have been isolated in a crystalline form, preserved over long periods, and photographed by the electron microscope. On analysis they turn out to be largely composed of the complex material also found in chromosomes of the cell-bodies which 'carry' the pattern of cell life and control the processes of self-regulation, reproduction, growth and adaptation—known as Deoxyribonucleic acid (D.N.A.) and Ribonucleic acid (R.N.A.).

Complicated systems of proteins and other organic materials have been manufactured under laboratory conditions and found to exhibit similar properties to very simple single-celled organisms, suggesting that the origin of life may have come about **via** such forms. At this level the words 'living' and 'dead' need re-definition or exclusion, in view of the emotional and unscientific overtones which they usually carry.

More recently still, the analysis of cell chromosomes by Watson and Crick in 1953, have shown the effectiveness of the physico-chemical explanation of living processes. D.N.A. molecules from chromosomes have been analyzed into component units, and in a number of cases the pattern in which certain key components are linked together in the enormous chain of the molecule has been shown to be decisive for cell duplication and development. In principle, therefore, it has been established that the processes of life can be explained in terms of physics and chemistry. It must only be a matter of time before the present area of explanation is widened still further. Once this has been done, simple kinds of living matter may eventually be synthesized under laboratory conditions. This may be still a very long time ahead—the complexity of the problems of synthesis are quite enormous—and the result is more likely to resemble an ameoba than a Frankenstein monster!

What then is life? The word is used in a wide range of contexts and with strong emotional overtones, but can anything consistent be said of its use in the biological sphere? What are the implications of such 'maps' of the word for those who use it in other spheres, particularly in religious language?

The issue posed by the earlier explanations of mechanists and vitalists or of those who, following Goethe, preferred organic concepts, is equally a problem for those interested in the religious use of the word. To say that life is the gift of God to his creatures is bound to appear to be in conflict with a statement that life is a mechanical system, or a natural force in certain things, or a property of whole organisms. In the past, too easy a reconciliation occurred between the religious notions of life and the notions of vitalism or organicism. If it is possible to reconcile any would-be scientific explanation, it is possible to reconcile all of them, including mechanistic or physico-chemical explanations. The prerequisite of any such reconciliation is the recognition that the **use** of the word 'life' in the religious context is an analogy with some ordinary experience. No possible reconciliation can occur unless it is admitted on the scientific side that such analogous usage is legitimate, and that other interpretations of phenomena—such as the behaviour of living organisms—are possible. Few would hold such an extreme position as to deny outright the possibility of analogies. After all, science itself uses a large number of models and analogies. The real problem today is whether we accept a scientific explanation as the final, normative explantion, compared with which other interpretations are secondary, marginal—or nonsense. At this point the claims of all religions, including Christianity, are in conflict with the assumptions of large numbers of scientists.

The Christian understanding of life, derived from the biblical point of view and confirmed by the continuing experience of the Christian community, is that it is the gift and expression of God. In the Christian view, living matter is not separated from non-living in any final way in creation. This is made clear not only by the story in Genesis and by the Incarnation; it is also a key element in St. Paul's teaching in the eighth chapter of his **Letter to the Romans**, in which the hope for the transformation of all creation is bound up with hope for man as a result of the Resurrection of Christ and the outpouring of his Spirit. When the word 'life' is applied to man, this implies that he is given responsible existence before God, and shares in some measure the divine attributes of freedom, care and creativity. This likeness to God, or being made in his 'image', is the ground of hope that man may share more fully in God's life—sometimes referred to as heaven or eternal life: a hope which is confirmed by the prophets and by Christ himself. The use of the word, even in the bible, is varied—almost poetic—at times. It should not be surprising if some find it difficult to adjust to a more systematic use of it in the sciences.

In spite of the anxieties and misgivings which have been aroused by recent biochemical and behaviourist investigations of living things, the only fundamental difficulty occurs when scientist or Christian claims an exclusive monopoly of truth for his particular perspective, or his use of words. Can such exclusiveness ever be justified?

3
The Nature of Man

The preceding discussion leads logically to a consideration of the impact of scientific thought on the understanding of human nature. Concern is currently felt about this impact in two areas: the description of the functions of the brain, and their implications for the notions of mind or freewill; and the science of psychology and its apparently complete, deterministic explanations of the inner aspects of human life.

Not only has the anatomical structure of the human brain been explored extensively in recent decades, but the physiological function of brain cells and whole areas of the brain have also been studied. The electrochemical processes by which nerve cells transmit 'messages' to and from the brain have yielded to patient chemical and physical investigation. The means of storing such impressions and of 'learning' responses are under intensive enquiry.

Models of the brain mechanisms have been proposed—particularly that of the computer, which behaves in some respects like a brain in storing and releasing messages under certain controlled stimuli. New computers are said to 'learn from experience', to be self-correcting, talking, thinking 'creatures'. It has been suggested that in future, computers will not only rival the speed and complexity of human thought, but also the whole range of human knowing and deciding. The 'Compleat Robot' is more than a pipe-dream, it is a confident prediction.

At the same time, it is pointed out that artificial intelligence is limited by the original programme given to the computer, and by its built-in lack of originality, subtlety or decisiveness in new situations beyond the programme. No computer has yet been made to decipher different kinds of handwriting, for example, or to write poetry or good music. There are several indications that the computer model for the brain has serious limitations as a model. After all the language of the human brain is not that of mathematics.

The question however is this: would a more adequate model of a brain, picturing the whole process of mental operations from a physical or chemical point of view, **reduce** man to a physico-chemical mechanism? To maintain such a view requires an equation between the concept of the brain and the idea of mind. It also supposes that a physico-chemical explanation is the **only** one to be offered for any entity or phenomenon—a view which has been questioned in the discussion of life above.

The relation of mind to brain is a constant issue in philosophy. We have noticed the influence of Descartes' view of the dualism of mind and body (including the brain), which led him and many after him to believe that mental and physical events are radically different (see p. 21). This provokes the questions: how do body and mind interact? Why are my mental processes dulled when my blood-sugar is low? What is the connection between the pain that I feel in my mind and the pin which punctures my skin? How can a thought issue in a physical act?

Professor Gilbert Ryle has suggested recently that mental concepts are really statements of dispositions to behave in certain ways. Observation of what people do when they say they think in certain fashions leads one to generalize mental notions in this way. The concepts of the mind are useful conventions—a kind of shorthand used by the individual to express what he may do.

This view of mind is not completely supported by those of the parallelist school, which maintains that mental experience is distinctive, and is always paralleled by a physical change. Thus a notion of blueness in the mind's eye is linked to some identifiable chemical or electrical process in the brain, or a dream experience corresponds to an electrical pattern which can be plotted on an electro-encephalograph.

Professor D. M. Mackay has advocated a two-language description of an event such as picking up a pencil to account for physiological and mental descriptions of it. One provides an observer's scientific description or explanation of electrical charges in nerve cells, and muscle stimulation and contraction, the other the actor's awareness of what he does. Physiological and 'mental' accounts represent the outside and inside of the same happening. Confusion occurs only when the categories, concepts and languages become mixed—as they tend to do in everyday speech. But this position raises the interesting philosophical questions: how can I know what is happening in other people's minds? If mental language is essentially the language of self-understanding, is it possible for two minds to communicate? How can we be sure that when two people refer to a mental event or experience—a pain for example—they mean the same thing? It is hardly surprising that so much attention should be paid by modern philosophers to the meaning of meaning, the nature of language, and the way words refer to things or events.

Another area in which considerable concern and controversy exists is that of psychology. For many devout people, psychological understanding is a very dangerous thing, likely to undermine faith and religion. Their fears have not been without foundation in the past. Sigmund Freud (1856–1939) wrote:

Religion is an attempt to get control over the sensory world, in which we are placed, by the wish-world, which we have developed inside us as a result of biological and psychological necessities.

Future of an Illusion, *p.* 57.

Elsewhere he said:

Religion is the universal obsessional neurosis of mankind.

Works *XXI, 44.*

In his analysis of the human psyche, Freud showed the vital importance of the early years of childhood. He stressed also the key role of sexuality in the human mind. On the assumption that what had happened in childhood was necessarily childish, and that the explanation of the origin of an idea was enough to exhaust it of any value, Freud maintained vigorously that religion was something which adults should grow out of—psychoanalysis had explained it away.

Like Newton and Darwin, Freud was a man of brilliance—if of some social eccentricity. He was an innovator of theories, and had the gift of communicating his ideas to a wide public. It is quite clear, however, that some of his assumptions, such as those mentioned above, were quite unjustified by scientific evidence or simple logic. His sweeping, dogmatic statements rested on somewhat limited clinical experience: most of his patients were people suffering from mental illness—hardly the most satisfactory sample for the establishment of criteria of normality. Nonetheless, Freud challenged the view that religious understanding and response had nothing to do with individual personal make-up, early experiences or human relationships. He also exposed the fact that religious and moral experience or practice are not always what they seem. From his time onwards, thoughtful people have been wary of making too much of individual religious experiences or practices. They have also recognized that psychology can say **something** about the way human beings think, feel, and behave, and how religious ideas, images and beliefs are developed, but that this is not all that there is to say about them.

Post-Freudian psychology has largely rejected Freud's anti-religious attitude. Carl Jung, in particular, has stressed the importance of religious ideas in his analysis, and even takes account of God as a 'psychological fact'. The other line of psychological study—the Behaviourist school—has taken up a neutral position about the reality of God. In its attempt to investigate observable facts of human behaviour, relating them with special concepts and explaining them by various theories, it can do no more than recognize religious patterns of behaviour among other kinds, and indicate some of the factors which shape them. Among Behaviourists there are those who hold that such explanations can in principle be final explanations of the phenomena, just as some Analysts maintain that understanding of the human mind can be achieved finally only in scientific terms. Once again, the issue is clear: can other interpretations have equal authority? In the last resort, which kind of understanding matters most? Meanwhile, the advance of modern psychology raises the further questions: can psychology tell us **how** the idea of God is developed? How does religion shape our lives? What is the influence of experience on religious understanding? Is any kind of experience specifically to be regarded as religious experience? If not, how is ordinary experience seen from a religious point of view?

4
Topics, Questions and Sources

Note: Questions for discussion can be found at the end of each section of the preceding chapter, on pages 69, 74 and 78.

Topics:

1
Will further research into D.N.A. and the genetic code mean that man will be able to control heredity, and develop special kinds of people?

2
Arrange a symposium with three speakers representing broadly differing views on life, for example, a biologist, a philosopher, and a priest (or member of a religious order).

3
Debate the motion: 'In the opinion of this house robots will rule the world'.

4
Investigate the meaning of the word **life** in *a* the Old Testament, *b* the New Testament, *c* Hinduism and *d* Japanese Buddhism. (Use a **Theological Word-Book of the Bible** *S.C.M.* 1950 and **World-Religions: A Dialogue**, R. N. Smart, *Pelican* 1695.)

Sources:

General—
Issues in Science and Religion I. G. Barbour *S.C.M.* 1966 *Chs.* 11 and 12.
Religion and Science J. S. Habgood *Mills & Boon* 1964.
Science and Religion: Conflict and Synthesis I. T. Ramsey *S.P.C.K.* 1964.

Evolution—
Evolution and Christian Belief David Lack *Methuen* 1957.
Evolution and Christians P. G. Fothergill *Longmans* 1961.
Scientific American—Charles Darwin Loren Eiseley 1956 *February issue.*

The Nature of Life—
History of Biology C. Singer *Schuman* 1950 *Ch.* 10.
What is life? E. Schrodinger *C.U.P.* 1944.
Biology and Personality Ed. I. T. Ramsey *Blackwell* 1965 *Chs.* 3 and 7.

The Nature of Man—
Man on his Nature C. Sherrington *C.U.P.* 1951.
The Physical Basis of Personality V. H. Mottram *Penguin* 1949.
Science Journal on **Machines Like Men.** October 1968.
Freud and the Post-Freudians J. A. C. Brown *Penguin* 1963.

Christianity in a Mechanistic Universe D. M. Mackay *I.V.F.* 1953.
The Biological Time Bomb G. R. Taylor *Thames & Hudson* 1968.

B
Explaining the Universe

1
The Conjectures of Cosmology

The issues posed by recent developments in cosmology are of a different kind from those which have been considered so far. They concern first the significance of man in the universe rather than his relation to the animal creation, and secondly the process of creation itself.

The use of large optical and radio telescopes like those at Mt. Palomar and Jodrell Bank has revealed that the material universe is unimaginably vast. Planetary distances considered by Copernicus, Brahe and Newton were a few hundred millions of miles, far enough to be almost beyond imagination. The improvement of optical telescopes increased the distances covered by observation to tens of billions of miles—a few light years—representing the distance away of the nearest stars similar to our sun. The most distant stars in our galaxy (containing some one hundred thousand million stars) are several thousand billion miles away. But now astronomical devices indicate that there are more than a hundred million galaxies and nebulae; the furthest at a distance from our system—as measured by the telescope at Mt. Palomar—extending to over one thousand eight hundred million light years! In fact, the more penetrating the instrument used, the more nebulae come into view, and in such a way that it seems the density of matter in the whole of space remains constant throughout the known universe. If to these unimaginable conclusions is added the fact that many of the stars are millions of times larger and hotter than our sun—and may have countless other planets in their solar systems—the overwhelming nature of modern astronomy can be understood.

The mind boggles at these dimensions, and more important, the claims made for the significance of man, a creature on a small planet in a minor solar system in but one of millions of galaxies, seem rather flat. The impression created by the results of modern astronomy is that man is only a fleeting speck in the universe, perhaps with many comparable forms in other solar systems, let alone our own. The claim of religion that God, the creator and upholder of this enormous system, is interested in man, cares for him and loves him, seems absurd. It appears to be hard enough for God to cope with the teeming millions of the human race on earth—quite apart from the rest of the universe and other existences as well! The ancient imagery of God sitting in the heavens above the circle of the earth has to go in the face of these facts, but what can possibly take its place? Man has eaten of the tree of knowledge—and most certainly lost his innocence here. The concept of God as creator has been exploded by the sheer size of the universe, but has it been destroyed? At the same time, has the possibility of religion, with its stress upon the relationship of man to God, disappeared?

In answer to this threat, which large numbers of sensitive people feel, it must be stated firmly that insignificance in size does not **logically** lead to insignificance in value. This is true of many things, such as a diamond ring, for example. The enlargement of the universe at one end is balanced by the discoveries of the electron microscope at the other. Man may be a speck of brief life-span in the material universe, but he is not by virtue of this **necessarily** unimportant or valueless to its maker. The significance of man, as Dr. Ian Ramsey has said, is not a spatiotemporal matter. Moreover, the very unfathomableness of the universe, far from destroying the element of mystery, in fact enhances it, pointing us to one of the characteristics of God the creator, who is incomprehensible, unfathomable and infinite in the religious sense. In other words, if the universe expands, then so can our idea of God!

While the newer pictures of cosmology break the old, seemingly infantile images of God, they do not shake in the least the basic idea of God behind those images as personal, ineffable, glorious in majesty and power. The new knowledge expands rather than explodes the older notions, refining our crude and limiting mental pictures, and increasing the sense of awe at the scope of the creation. The mystery of God's concern for each human person, and the wonder of the Incarnation are heightened by this new knowledge, if they are received by men of faith. In the nature of the case they cannot be disproved by it. The 'problem' of other human types on other solar systems then becomes merely a problem of human pride. The Christian answer must be—so what? and why not? God is not limited, by time, space, matter, energy or human self-concern.

Modern cosmological theory poses another issue to religion in relation to the nature of creation. It has been suggested that the explanation of the darkness of the sky at night is to be found in the expansion of the observable universe. Using the spectroscope to form a spectrum of colours from the light of distant nebulae, M. Humason showed that instead of certain dark bands in the red region of the spectrum appearing in the position they normally have in spectra of light from our sun, they are displaced towards the shorter wavelength. E. P. Hubble explained this with the theory that the distant sources of light were moving away, the shift being an optical Döppler effect, like the change in note (sound wave-length) of a fast moving vehicle as it passes a stationary observer. Assuming the operation of the normal laws of physics, this means that the distant galaxies and nebulae are moving away from us—the universe is expanding on a vast scale. This theory, which is widely accepted, has led cosmologists to speculate on the possibility that the whole material universe had a single origin. On the basis of the rate of expansion calculated by Hubble, it has been suggested that at a moment of time some 7,000 million years ago, the whole of the matter in the universe would have been concentrated into one enormously dense unit, which then exploded.

Other speculations than this 'big bang' idea have been aired to account for the evidence. Among these is the 'continuous creation' hypothesis associated with the name of Professor Fred Hoyle. Accepting the conclusion that the universe is expanding, this hypothesis maintains that matter is being continually created to maintain the constant density of matter in space as the galaxies recede over the observable horizon. This suggestion requires that an atom of hydrogen originates in a space the size of St. Paul's Cathedral in every thousand years. Hoyle suggested that it might even be a property of space itself to create matter, and thus the notion of a divine creator would become superfluous!

These ideas about creation remain hypotheses, as yet neither decisively confirmed or disproved. It is likely that other conjectures will continue to appear as the exploration of cosmic ray sources and radio stars goes on. But the implications for religious interpretations of creation have been seized upon eagerly by Christian and agnostic. Hoyle's earlier view that God was now rendered totally unnecessary in particular raised the dust of controversy once again. He has modified it since.

In point of fact, these hypotheses present us with the same basic problem that has been met before. Even if science can give us an explanation of observed phenomena, is this the only or the complete explanation? Even if Hoyle's hypothesis of continual creation proved to fit all the facts and to stand up to every kind of test, would God be rendered obsolete? The answer depends upon one's prior conviction about the nature of reality, and the Christian will confidently affirm that God is quite capable of creating his universe with a bang or a whimper—through 'creative space'. Neither hypothesis touches the theological truth of which the doctrine of creation speaks. Neither explanation can exhaust the meaning of the time-long phrases 'Creator of heaven and earth' and 'Maker of all things, visible and invisible'. The explanations of cosmologists consider material things, those of theology are about the spiritual reality which underlies them.

2
Nineteenth Century Physics

A distinguished mathematical physicist once remarked that there were two kinds of scientific imagination: one which worked in the terms of mathematics, with algebra and geometry; and the other which worked in terms of pictures, by means of visual images. If Galileo and Newton are typical of the first kind of imagination, Michael Faraday is representative of the second.

Faraday was born in 1791 at Newington, in Surrey. His father was a blacksmith, but a man who suffered bad health and whose resources were meagre. The Faraday family belonged to the small Christian sect known as the Sandemanians; an earnest, fundamentalist group, membership of which contributed much to Faraday's straightforward, humble approach to his work in later years. At the age of 13, Faraday took a job as an errand boy in London, and went on to learn the trade of book-binding. Meanwhile his interest was aroused in the 'new' phenomena of electricity, about which so little was then known. He saved hard, bought books and equipment, and began to perform simple experiments which he carefully recorded.

Faraday managed to attend a course of lectures on chemistry given at the Royal Institution by Sir Humphry Davy. He made extensive notes of the course, and having grown dissatisfied with his career in the book trade, sent them to Sir Humphry asking for a job. Davy employed him as a laboratory assistant and he soon showed his considerable skill in practical work. As a result

Faraday came into contact with the scientific élite of the day—Ampère, Gay-Lussac, Clement and others. He toured Europe with Davy and built up considerable practical and theoretical knowledge over a number of years.

In 1819, Oersted discovered the existence of a relationship between an electric current and magnetism. A compass-needle was deflected from its normal position by passing a current through a wire held over it. Faraday repeated this and other experiments, and with some ingenuity showed that a force large enough to turn a magnet could be generated by passing a strong current down a coil of wire. He also discovered that the force rotated the magnet into a particular direction. To explain this, Faraday postulated 'lines' or 'tubes' of force around the electrified wire, circling in a plane at right angles to it, stretched into space, and returned to the wire in the form of a loop. Thus a visual representation of this intriguing phenomenon was given and used to explain electromagnetism.

Faraday went on to show that a given quantity of electricity produced a constant force, and that variations in the current were accompanied by corresponding variations in the magnetic forces associated with them. In this work, the origin of the electric motor and the dynamo are to be found.

By other experiments in electrolysis—passing an electric current through a chemical solution—Faraday showed that a given quantity of current produced equivalent quantities of components in the form of deposits or evolved gases with the substances he used. Thus in two basic practical experiments, he laid the foundations for two major areas of modern science, and provided vital evidence for the culminating theory of classical physics. Having succeeded Davy as Director of the Royal Institution, Faraday remained, a devout and much admired man, until his death in 1867.

James Clerk Maxwell was the product of a very different background—rich, well-educated, urbane. He was more typical of the earlier, leisured members of the Royal Society of Newton's day. It fell to him to develop Faraday's data into a comprehensive theory of classical physics. Extending Dalton's atomic theory, Maxwell first proposed physical properties for atoms to explain the phenomena of heat and the diffusion of gases. This theory, published in 1866, added support to the atomic classification of chemical properties, which was proving so fruitful as the elements of the Periodic Table were isolated one by one.

Maxwell next turned his attention to the phenomena which Faraday had discovered. He proposed that the explanation for them could be found in 'particles' of electricity associated with the physical properties of particles already used as the basis of physical theory. On the notion of electromagnetic lines of force, Maxwell used Descartes' idea of aether—the 'substance' which was thought to fill space and through which particles could act and react on each other at a distance. Electric 'particles' exerted a force on the aether, said Maxwell, distorting this invisible substance, setting up stresses in it and thus containing energy. It was this which was being mapped by Faraday's tubes of electromagnetic force.

Using Euler's mathematical analysis of hydraulics, Maxwell elaborated his 'field of force' theory, and produced mathematical equations which allow wave motion. The velocity of this wave is related to the electromagnetic properties of aether. He was able to determine these properties quite simply in the laboratory, and found that his result compared well with the value for the speed of light obtained by Foucault and Fizeau! He then suggested that the wave motion of light was a transverse oscillation in the same medium (aether) as the motion of electromagnetic force. He concluded that his theory of fields operating through this medium was the basic explanation of the phenomena of light and magnetism.

The simplicity of Maxwell's explanation was overwhelming. It seemed that the final explanation of physical, chemical, light, heat and electromagnetic phenomena had been reached, and that the actual picture of the physical universe—in its fine structure at least—could be drawn. By 1890, the vast majority of scientists accepted the theory and the classical nineteenth century synthesis seemed complete. But at this very moment the chemists Wilhelm Ostwald and F. A. Kekulé began to voice doubts about the finality of the model. Jean Perrin had begun his work on Brownian Movement—the motion of fine particles suspended in a liquid—which was to be the experimental basis of Einstein's first major achievement. J. J. Thompson was setting about the discovery of the electron in Cambridge. The classical, mechanistic picture was about to be exposed for what it was: a useful, but strictly limited model.

3
The Breakdown of Literal Pictures

Classical physics, brought to a final moment of glory and comprehensiveness by Clerk Maxwell's electromagnetic theory seemed to achieve the final explanation of the structure of the universe. The basic units of matter and energy were defined, described, grasped by billiard-ball, wave and stress concepts in such a way that the secret of the whole universe appeared to be contained in the elementary units of which it was made.

The theory was **deterministic**—that is to say that the future state of the systems described by the theory (such as the change of gas to liquid under pressure or sum total of energy in the universe) could in principle be predicted from information about the present state of things in the system. It was **reductionistic**—all laws could apparently be derived from the mechanics of particles, and laws governing particles, fields of force and wave motion. This applied to solids, liquids or gases; to heat, light and electricity, and to gravitation or chemical affinity. Ideally, explanations of all phenomena could be reduced to quantitative changes in simple units which remained basically unaltered in character. The theory was **literalistic**—that is to say the notions of particles or 'field of force' were taken to be exact representations of what was there in material reality. However experiments were performed, the objects under investigation would appear to be the same. Thus the attitude taken towards the everyday world of books, tables and chairs, which we assume are there for everyone to see or feel, was applied to the structure of the whole universe.

Certain new experiments were conducted at the end of the nineteenth century, such as Max Planck's investigation of the spectrum from a hot object, and Perrin's work on Brownian motion—the movement of tiny particles of pollen suspended in a solution—from which it became clear that the classical physics could not give a satisfactory answer for all the observations. From measurements of the continuous spectrum with its unexpected dark lines Planck was able to deduce that the atoms of the radiant object must vibrate with certain specific energy values. In the period 1890 to 1900 several remarkable new discoveries were made of electrons, X-rays, and radioactivity. In 1913 Niels Bohr produced a model of the atom with a heavy central nucleus and electrons orbiting around it—a miniature solar system. He then showed that the electrons could only stay in their orbits if certain discrete energy levels were possible. In the theory of the photoelectric effect, put forward by Einstein in 1905, Planck's explanation was extended by the suggestion that radiation energy existed in distinct packets—such packets or units of energy being called photons. A beam of light or X-ray was thus described as a succession of energy units, or photons, rather than a smooth flow of wave energy. Such units could behave as if they were waves and also as if they were particles having an impact on a surface and giving off energy.

From this point, the new electromagnetic theory was to undermine mechanical theory as the lowest level of explanation in physics and chemistry. In 1924 de Broglie advanced the idea that matter—as well as light—might have a 'double character'. Using a mathematical analogy, he suggested that electrons might behave like waves. Schrödinger then worked out the necessary mathematical equations to describe such matter-waves, showing that there was a **probability** that a wave system of the simple hydrogen atom would behave like a single, localized electron in accordance with the size or amplitude of the wave at any moment. He then indicated the kind of experimental evidence which would support this hypothesis. In 1927 Davisson and Germer provided the experimental evidence for this, showing how a beam of electrons reflected from the surface of a nickel crystal produced diffraction patterns like those obtained using X-rays of similar energy.

Next, Werner Heisenberg, at the age of twenty-five, showed from mathematical calculations based on the Quantum Theory that there was a theoretical limit to the accuracy with which it was possible to measure certain pairs of variables concerning the electron. If the position of an electron was pin-pointed in a moment of time, he showed that it was impossible to determine its momentum; if its momentum or speed was measured, it was impossible to locate it. This is his **Principle of Indeterminacy.**

Just as in earlier times, scientists had found it hard to accept a change of concept, so too in this case, many found the change from classical physics with its billiard-ball entities and miniature atomic solar systems to the more abstract mathematical notions of Heisenberg and Schrödinger difficult to make. Lord Rutherford, a remarkably open-minded and ingenious experimental worker, could never quite accept it. Albert Einstein, the mathematical theoretician, did not believe that the new mathematical model was the last word in explanation, suggesting that it was only a temporary stage which would be superseded as it became possible to make better measurements and develop better visual models. Recent theoretical physics has, however, been based on the Heisenberg–Schrödinger theories of 1926–27. They have led to the postulation and 'identification' of the scores of sub-particles-cum-wave-entities: positrons, neutrons, and many more. The sheer number of these real or hypothetical entities, identified inside or outside the atom, has suggested to some that there may possibly be a new theory one day to simplify the picture. But it is clear that theory is at least as important as experiment in the sub-atomic world of modern physics.

The wave-particle dualism of electrons corresponds to the apparent behaviour of light in different experiments. In some situations, light behaves as though it were a wave—in interference, for example. In others, like the photoelectric cell, it behaves as if it were a stream of particles. Different explanations are needed for different circumstances. Bohr termed this the **Principle of Complementarity**. He also showed the necessity of simple models in designing experiments and describing results, however inadequate they may be, and also pointed to the fact that the hypotheses and experimental methods of physics determine the **kind** of fact which may be observed. In other words, the method of observation is involved in such 'facts', and belief in the absolute realism of scientific models cannot be sustained.

Thus modern physics has lost the power of realistic picture-making, at least at the sub-atomic level. It recognizes the role of the observer, and the limitations of experimental methods. The conclusion to be drawn is that the material world is indeed mysterious and in some ways eludes our grasp and understanding. The implications of work in the field of astronomy are exactly the same.

4

Unsuccessful Reconciliations

The considerable mental shift in modern physical science away from the 'final explanations' of classical physics in the nineteenth century has important consequences. It has raised philosophical questions about the nature of scientific knowledge. It has exposed the limitations of scientific 'facts' and theories, and the inadequacies—as well as the crucial importance—of models and analogies. The present state of physics underlines the symbolic nature of both quantitative and qualitative terms. Most important, our lack of precise models or pictures of the most fundamental level suggests that we are not dealing with a basic system of matter and energy in which the future behaviour of the entities can be rigidly determined by what has gone before. The modern view of this level of things is that only time can tell which of the various possibilities or potentialities will come to pass. Such an open-ended view requires a new attitude to time, even to history, as unique, unpredictable (beyond certain narrow limits) and unrepeatable. This is in marked contrast to the deterministic processes of classical science, typified by the theory of the dying universe, slowly loosing its energy, derived from the Second Law of Thermodynamics.

Various interpretations of the notion of uncertainty are currently held. Einstein believed that it reflected our present lack of knowledge. Bohr thought that it indicated a limit to experiment or human concepts which is quite fundamental. Heisenberg and Margeneau say that indeterminacy is a feature of nature itself, to be accepted as a fact in its own right.

At this point the philosophical and religious problems become acute. W. G. Pollard, the Director of the Oak Ridge Atomic Research Institution, Tennessee, who is also an Episcopalian priest, has suggested that the fundamental element of indeterminacy in the sub-atomic world is the 'place' where God's providential control of the universe is exercised. He can at this point 'influence' events without acting as a physical force, and yet cannot be scientifically discerned.

It would seem to be the case, however, that the deterministic system of classical physics has to be met on other grounds if proper place is to be given to the notions of freedom, and the realities of religion. To the physicist, the opposite of determinism is not freedom but chance. The idea of freedom carries with it the notion of choice, even conscious choice. Nobody could reasonably suggest that the behaviour of an electron is a matter of such choice. Even if there is at this level a break in the pattern of cause and effect which serves as an effective working assumption throughout the rest of science, all we would have would be randomness or chaos, not freedom. To introduce supernatural influences at this point, and at this point only, is to return to the days of alchemy. Unless God is reduced to a physical reality—a 'God of the gaps'—his interaction with the material world through chance or anything else cannot be located specifically here. Similarly, the notion of freedom of choice belonging to the realm of mind or personality cannot be 'placed' within the interstices of the atom.

This is an illustration of the mixing of categories, discussed before. Supernatural concepts refer to the supernatural realm. The spiritual world cannot be confused with the material without reviving a vulnerable notion of a 'God of the gaps'. In the same way, concepts like that of human freedom belong logically to the 'language-map' of personal discourse, not to that of any science. Even though a deterministic situation may be disclosed by a scientific explanation of human thought and action, Dr. Johnson's remark, 'Sir, I know I am free, and there's an end on't', is the complete answer at the personal level. Likewise, however precise a description of any phenomenon—its causes and effects—may be given by the sciences, the interpretation of the event in terms of God's action is untouched. Such an interpretation has to be examined at the theological rather than the scientific level. It might well have been possible to give a complete scientific description of the exodus, the miracles of Jesus, or even the resurrection, but the significance of these events as the acts of God has to be assessed in the context of history by **theological** criteria. Is this consistent with God's nature? What is the religious meaning attached to the event? Can the disclosure of God's will and nature in the happening be sustained by what happened—however it is understood? Questions like this enable Christians to distinguish revelation from fantasy, and authentic miracle from magic.

Other writers have inferred from their reading of modern physical theory that matter is now to be understood as 'insubstantial', consisting of empty space and probability waves of energy or matter. Since matter can be dematerialized into energy, as in atomic fission, the basic nature of solid stuff is therefore supposed to be immaterial. They have suggested that this is evidence for the active, creative Spirit of God in the physical realm.

Again this has been too hasty a conclusion. The fundamental facts of sub-atomic physics may be more accurately represented by the recent theories than by the billiard-ball notions of nineteenth-century physics, but the newer concepts of energy, electric fields and probability waves are no less **physical** than Dalton's atoms. They are measured by specific physical experiments. To suggest that matter is not really solid is nonsense, if it is thereby implied that the table no longer supports the book, or if you bump into a chair you are likely not to feel anything as you pass through it!

Here, too, there is a confusion of terms. The word 'insubstantial' is used in different ways in physics and theology. In one case it refers to the condition of matter at the lowest level. In the other, it is used analogically of God, the spiritual reality. Its dual usage is confusing until the context is carefully examined, when the absurdity of equating the two kinds of insubstantiality becomes clear.

In fact, this claim to find evidence for God at the sub-atomic level is an attempt—all too often made—to get round the difference between God and his creation. If it had been successful, it would have eliminated faith by providing a 'proof' of God's existence by scientific means. But God would then have ceased to be God as Christians claim him to be—ineffable, incomprehensible, infinite, trancendent. This poses the continuing problem facing modern man, of finding the right distinction and the right relationship between the spiritual and the material realms; and between the languages of science and faith.

5
Topics, Questions and Sources

1
Compare a modern cosmological theory of creation with that of St. Thomas Aquinas. (Use **The Nature of the Universe**, F. Hoyle *Penguin* 1964, and **Aquinas**, F. C. Coppleston *Penguin* 1959).

2
Is it possible in principle for there to be one kind of explanation—*e.g.* a physical one—for all kinds of phenomena?

3
How does the latest exploration of space by manned spacecraft help in understanding the nature and purpose of the universe?

4
Debate the motion: 'In the opinion of this house, the principle of indeterminacy ends the quest of modern physics'.

Sources:

Issues in Science and Religion I. G. Barbour *S.C.M.* 1966 *Ch.* 10.
Science and Christian Belief C. A. Coulson *Fontana* 1964.
The Atom and Its Nucleus G. Gamow *Prentice-Hall* 1961.
The Nature of the Universe F. Hoyle *Penguin* 1960.
The Modern Universe R. A. Lyttleton *Hodder* 1957.
The Maker of Heaven and Earth L. Gilkey *Doubleday PB* 1959.

C

Language Games

1

The Final Explanation?

This discussion of issues in the debate between science and religion has constantly returned to the contrast between scientific and religious language. The main issue in the conflict between some scientists and some religious people rests on the question of the finality of either explanation. There are those in each camp who maintain that their pattern of explanation is the only valid one, any other being either meaningless or mere conjecture.

A good measure of the unease still felt by devout people at the advance of scientific explanation has arisen because they seem to assume that science **does** in some way have the last word, and there is a risk that everything will be explained away or reduced to a scientific analysis. Certainly a large number of scientists subscribe to this view. Here is a typical recent statement by an eminent astronomer, Professor R. A. Lyttleton:

> . . . . *the steadily emerging conclusion, which science itself is forcing upon us*, (is) *that almost all that has hitherto passed for knowledge and understanding in other fields, such as political*

theory, philosophy, and ethics, to mention only a few, will eventually be relegated to the level of mere rationalisation: that is 'associated explanations' having no real validity.

The Modern Universe, *p. iv*, 1956.

If this is the case, then most certainly religious, moral and aesthetic understanding will be reduced to some appropriate 'scientific' form, and the game will be up!

On the other side there are still those—a small number—who react strongly against scientific explanations of biblical events, human behaviour, life or the human brain. Such explanations seem to them blasphemous, or at best passing theories which are likely to be replaced in the near future. It appears that neither extreme can possibly meet: fundamentalism about religious understanding and about scientific explanation are mutually exclusive, and tend to be linked with a psychological need for a single pattern of explanation as a form of intellectual security.

The powerful belief in science—'scientism' to give it a proper name—goes hand in hand with the philosophy of logical positivism. This is a philosophy of meaning originating among a group of philosophers and scientists who met in Vienna in the 1920's known as the 'Vienna Circle'. It was popularized in this country in the 1930's and 1940's by Professor A. J. Ayer and his school. Ayer's exposition of it is to be found in his influential book **Language, Truth and Logic**, though he has since modified his position somewhat as a result of various criticisms.

Essentially, logical positivism is concerned with the meaning of propositions. Its main tenet is that the meaning of a statement is given by the way in which it is verified or tested empirically. A scientific proposition such as 'water boils at 100°C under normal pressure' is meaningful because it is possible to test it experimentally. A religious statement such as 'God is love' is meaningless because there is no comparable experimental test. Similarly moral statements, 'that was a good thing to do', and aesthetic statements, 'this is a beautiful picture', while they serve to express emotion, approval, subjective attitudes or cultural standards, are logically unverifiable and therefore (by definition) meaningless in the strict sense of the term.

It was not long before philosophers pointed out that the main proposition of logical positivism itself was meaningless according to its own definition of meaning. The proposition 'the meaning of a statement is given by the way in which it is verified' cannot be tested empirically! The assumptions of this way of thinking were thus made plain. All the same, such was the influence of this philosophical approach, and such were its connections with science and technology, that many people have been convinced by it in a vague kind of way. In turn this has led pious people to retreat from any conscious dialogue with philosophers or scientists, and to rest content with strident assertions of traditional dogmas or personal faith, neither of which cut much ice outside Christian or other religious communities. In fact, many Christians have paid little attention to the problem of meaning of religious statements, being quite content to think of them as historical 'facts' or expressions of personal conviction and feeling.

2
The Logic of Language

The problem of the meaning of statements is as old as philosophy itself. Plato and Aristotle devoted much of their intellectual effort to the task of clarifying the meaning of words and propositions. In the tradition of modern philosophy, the English Empiricists—from Locke, Hume, Berkeley to Russell and Moore—have all been concerned with the meaning of statements in relation to the external world or logical necessity. The most refined exposition of this tradition was published in 1919 by a young Austrian engineer-turned-philosopher. It was called the **Tractatus Logico-Philosophicus**; its author's name was Ludwig Wittgenstein (1889–1951).

This remarkable work of a young man expounded a picture theory of language. He argued that words always name objects, that language, when it is analyzed carefully, consists only of formal propositions, and that these propositions 'picture' states of affairs. They do this by having a logical structure of representing signs of the combination of objects which makes up the state of affairs. To give an example: a gramophone record, the musical idea, the sound waves, are all related logically to each other in the 'picturing' fashion. This relationship holds between language and the world of objects.

It was through the study of the **Tractatus** that the members of the Vienna Circle came to formulate their particular theory of meaning. Moritz Schlick, its leader, came to the conclusion that science should be defined as the 'pursuit of truth' and philosophy as the 'pursuit of meaning'.

While the philosophy of logical positivism was dominating the scene, Wittgenstein—by now an eccentric, passionate Professor of Philosophy at Cambridge—was developing a new approach. Not content with being the master mind behind one major phase of understanding language, he became the originator of its second phase—an achievement which places him among the giants of philosophy.

At the end of his **Tractatus**, Wittgenstein concluded that the basic propositions on which his interpretation of language—giving a logical picture of the world of objects—rested, did not belong to that world. He clearly recognized that there were problems beyond the limits of science, thought or language about the world and about language. This area he termed the **mystical**, of which he spoke in sentences like these:

> *The sense of the world must lie outside the world. (6.41)*
> *It is not* **how** *things are in the world that is mystical, but* **that** *it exists. (6.44)*
> *Propositions can express nothing of what is higher. It is clear that ethics cannot be put into words. (6.42 and 4. 21)*

Tractatus Logico-Philosophicus

In his second major work, the **Philosophical Investigations**, published after his death in 1953, Wittgenstein moved away from his single-minded view of the nature of language. He came to acknowledge that there is not one single logic controlling the use of words and names, but as many logics as there are ways of thinking and speaking. In this Wittgenstein made a completely new departure in philosophical understanding, of which his successors have been quick to take advantage.

A word, said Wittgenstein, is not always a name for something; it can be used as a name but it can be used in numerous other ways, like the word 'fire!' The meaning of a word can be the thing to which it refers, but just as often the meaning is given by the way in which the word is used in language—what Wittgenstein called the language game. This is reflected in the odd usage of language in Lewis Carroll's work: 'When I use a word,' Humpty Dumpty said, 'it means just what I choose it to mean—neither more nor less.' Clarity of understanding—the task of philosophy—consists in looking at language to see how it is being used, and grasping the particular logic being employed. In a famous passage, Wittgenstein wrote:

> *How many kinds of sentences are there? Say assertion, question and command?—There are* **countless** *kinds: countless different kinds of use of what we call 'symbols', 'words', 'sentences'.*
>
> **Philosophical Investigations** *23.*

This way of looking at language has stimulated philosophers to make logical 'maps' of different uses of language, and clarified innumerable problems which arise when the logic of one usage has been confused with that of another. Meanwhile it is accepted that the logic of scientific language is akin to that outlined by Wittgenstein in the **Tractatus**. This has led to a clear acceptance of the limitations inherent in scientific explanation—of understanding the world of facts, that is of combinations of objects having distinct names about which it is possible to construct language-pictures in the form of propositional statements.

Under these important philosophical influences, attempts are going on to reconsider the precise nature and limits of scientific understanding, the status of religious language, and possible connections between them. The outlining of logical maps—'linguistic analysis' as it is called—has shown how certain words crop up constantly in several language games, and behave in a well-defined manner as logical constants. Words such as the personal pronouns 'I', 'You' and 'We' occur in many of the language-maps, and serve to unite the different language games, anchoring them in the understanding or use of the human subject. Professor P. F. Strawson has shown how personal terms of this kind unite mental and physical descriptions of the same event—such as picking up a pencil to which we referred earlier (page 76). These words provide a practical way of speaking about our own and other people's actions, thoughts, feelings or desires without troubling to analyze them according to the formal language-maps of mind or physiology. The important point is that formal mental descriptions and formal scientific explanations are special, abstract ways of understanding: only personal categories retain the commonsense, total view of what goes on, which is readily applied to ourselves or other people.

Professor Michael Polyani has pointed out that scientific objectivity is another name for the personal involvement of the scientist with an intention to make his work universally testable. In other words, scientific objectivity is the agreement of other scientists!

Dr. Ian Ramsey has taken this kind of language analysis a stage further in relation to religious thought. He has stressed that the third-person, objective language of science conceals the fact that it is 'observer' language, and implies the mind, imagination, outlook, technique and understanding of the scientist as a person—though the language is constructed in such a way as to minimize this subjective element as much as possible.

Dr. Ramsey has further shown how the language of religion is akin to personal language in its logical behaviour. If the doctor tells me that I have a pulse rate of 95, such a statement presumes that I exist, even though the statement 'I exist' does not require me to have a frantic pulse rate. Similarly, the statement 'Water boils at 100°C at normal pressure'—or **any other scientific statement**—may be said to presuppose the theological statement that 'God exists', although nothing verifiable may be deduced from that bare theological statement by itself. Theological statements, though similar to scientific ones in their use of models, analogies, and clear sets of rules, differ from them in being untestable, and behaving like personal statements. In a theological statement or story, attention is focussed on the personally significant features of the situation, not on the scientific or merely historical elements. The biblical account of the exodus from Egypt, for example, gives a heightened description of winds blowing, waters piled up in heaps and Egyptian chariots being washed away. The significant element in the story, long recognized in the Jewish Passover ceremony, is the providential rescue of his people by God. The historical facts are necessary to the episode and its meaning, but the theology is about God's relation with his chosen people. That part of the language is personal and unverifiable in the scientific sense. Only if it could be shown without doubt that **nothing** historical took place at all would the theological statement fail. Thus the historical and scientific details are subject to historical or scientific criticism: the theological statement, while it rests upon the necessary minimum of event, is subject to verification in the logic of theology only.

Ramsey believes that personal and theological categories can provide an outline for a total language-map, into which the separate logical languages of ethics, history and science can be placed. Not everyone would agree with this suggestion at this stage, and a good deal more work on the interrelationship between theological and other languages is required.

In the light of this trend of thought, we may say that a scientific explanation of all phenomena is in principle possible. At the same time we must roundly deny that such explanations are final. Likewise, we must refute the suggestion that scientific knowledge is the only kind of knowledge. Important as scientific explanations are, and useful as they can be, they cannot touch those matters which constitute the most important things in life: freedom of choice, love, faith and hope. These things are the stuff of religion, of which Humanism, Hinduism, Judaisam and Christianity are all examples.

3
The Language of Science

The community aspects of science have already been noted in the brief discussion in Section One. The character of modern science has been formed by the sharing of ideas and information across national, racial and ideological barriers, and by the testing of conclusions by other workers. By this means, the scientific community has developed certain common attitudes which are being shared to an increasing extent with the community at large. It has certain basic assumptions—such as the belief in the uniformity of nature, and the relationship of cause and effect in observable phenomena—which can never be scientifically or logically proved.

The scientific community has also developed its own language—or more strictly, languages, since every science has its peculiar terms and conventions, even though **logically** the languages of science keep to the same basic rules. The sciences use symbols to refer to observed entities and quantities. The method of science is to select certain features of a situation and discuss them rather than the whole system all at once. This isolation of topics is an extremely important feature of the scientific approach to problems. Mathematical notations are used increasingly, indicating the abstract nature of scientific theory, as well as the attempt to picture the relationship of observed phenomena as precisely as possible. The story told in Section One shows the way in which scientific thought uses analogies and models. (An analogy being a well defined correspondence or similarity between two areas of experience; a model being a pictorial representation of one system drawn from another system.) One of the best-known analogies in scientific history is the snake of F. A. Kekulé, the German chemist. Puzzling over the formula for benzene, Kekulé happened to dream of a snake which had taken its tail in its mouth, forming a ring (*figure* 2). He awoke and realized this was analogous to the structure of the benzene molecule. A familiar model in physics is the conception of light travelling in straight lines. A geometrical figure can be drawn to model the light source, object and shadow, which when related give a straight line path for light.

Models and analogies have been used since science began among the Greeks. Vivid examples are to be found in the writings of Democritus or Ptolemy, as well as in those of Sir Lawrence Bragg and Drs. James Watson and Francis Crick (*figure* 3). Such devices suggest new experiments for investigation, although they are themselves always limited and cannot be taken literally or overextended. The literalist period of science now belongs to the past.

Dr. Ian Ramsey, in his small but important book, **Models and Mystery** (1964), has indicated how scientific models are used to disclose the situation which is there in reality by representing selected features of the situation which can be treated experimentally. In any such system, certain features are taken as constants, such as the speed of light, or the molecule. The explanation is then built around these constants according to the rules and relationships required by the model—until such time as the experimental evidence calls into question some feature of the model, or even the constants. And always the observer is involved.

As might be expected, scientists do not always agree among themselves about the relation of their language to reality. Professor Ian Barbour has distinguished four separate views about this relationship: the Positivist, Instrumentalist, Idealist, and Realist.

The **Positivist** view stresses the key importance of observed data. The language of science is therefore primarily mathematical

figure 2

Kekule's Dream

The Benzine Ring (C6H6)

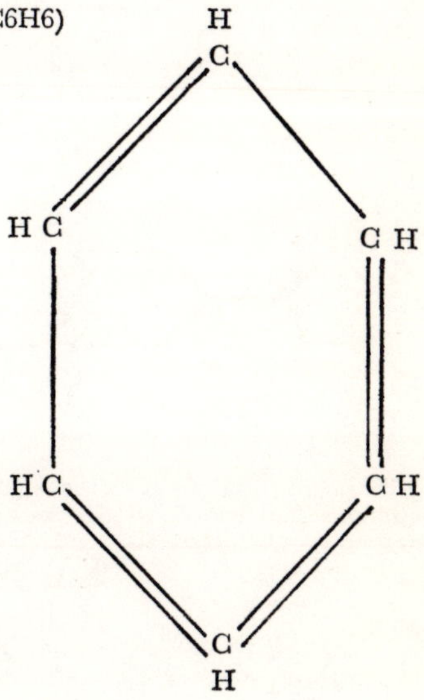

figure 3 **Two-dimensional model of D.N.A. chain**
(after Watson and Crick 1953)

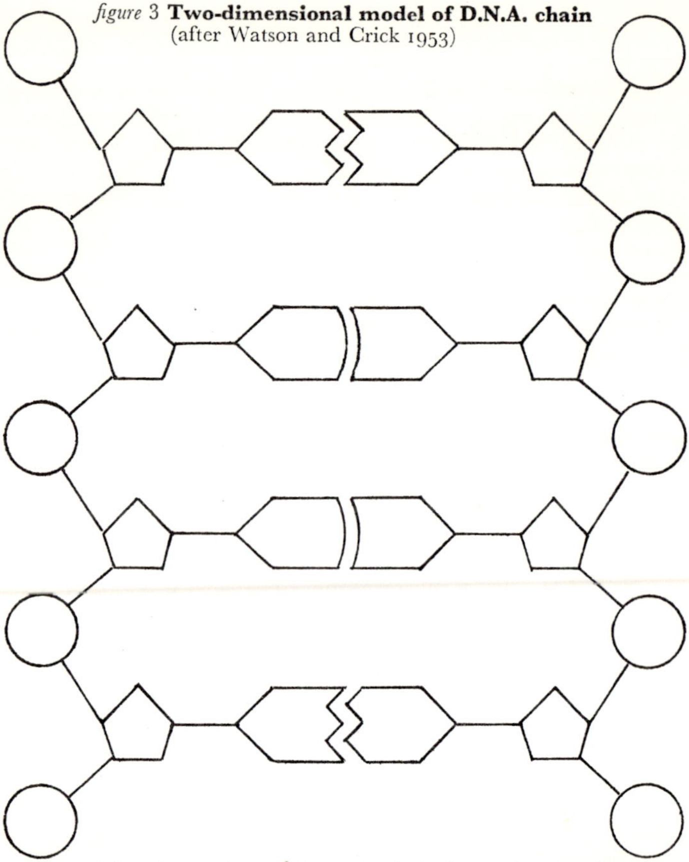

X-ray diffraction analysis of the molecule indicates that the chain is twisted into a double helix, which is pictured by a twist to the 'ladder'.

Key

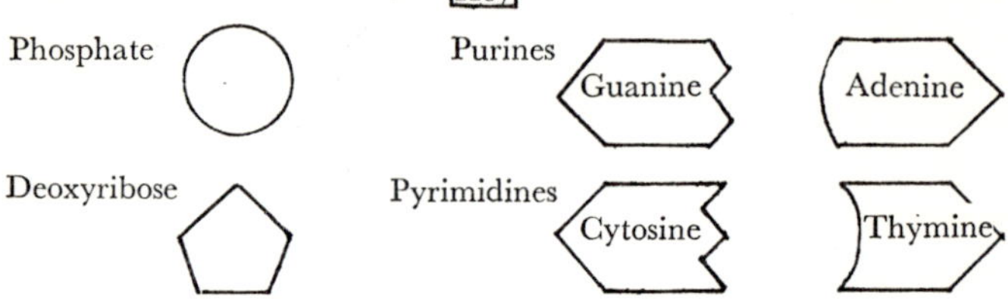

and quantitative. Scientific theories or models are then regarded as ways of classifying observations, and scientific concepts are ways of summarizing or representing facts in a simple manner—but neither are real in themselves.

The **Instrumentalist** view is that scientific laws are rules which enable the investigator to find his way about. Theories are techniques for making predictions which can be tested experimentally. Scientific language is thus a tool for setting up further investigations or putting data to use. Concepts are related to observed data on this view, but they are not simply observations. A molecule is a concept which serves as a useful instrument or handle for the investigation (and control) of normal units of matter—but nobody has ever seen one!

The **Idealist** view stresses the theoretical aspects of science, holding that the mind imposes its patterns on the chaotic mass of sense-data available. Sir Arthur Eddington, who subscribed to this view, went so far as to say that the fundamental laws of physics are wholly subjective, depending on the way that scientists have come to look at the universe. For such a view, scientific language is a means of rationalization achieved by reflection upon the experiences of scientists.

Finally, the **Realist** view is that there is a link between the pattern of scientific explanation and what is actually there in nature. Science is a process of discovery, not one of invention. But what is real is that which is understood rather than that which may be observed, as the positivists suggest. Electrons are real—though their existence can only be inferred from their effects. The language of science, on this view, is descriptive, even though it may be selective, partial and imperfect. Realists accept the possibility that scientific language may be limited by the means of investigation used, the concepts available and the imagination of the human mind.

Far from presenting a completely clear view of scientific knowledge to the world, the present state of the philosophy of science contains several different accounts of what science is, and what scientific language does. This may not alarm working scientists very much within their specialist descriptions, or the technologists who make such effective use of their findings, but it should remind us that there are many problems about science itself which have yet to be resolved.

A point about the language of science made by Lord Ritchie Calder and others is that the proliferation and fragmentation of scientific study has increased the use of specialized language to the point where communication in meaningful terms, even between scientists, is becoming impossible. For example, at a symposium on DNA, the chairman, a solid state physicist, said 'What **are** they talking about?' A chemist said 'I wish I could grasp this genetics stuff.' A geneticist said 'This crystallography has got me beat!'

The degree of human involvement in the patterns and methods of scientific understanding may be in dispute, but not the fact of such involvement. We can see, therefore, how it is possible to place science alongside art, history, music and politics as one of the humanities. From the glance at the history of science in Section One, we have seen how scientific thought has been influenced by dominant ideas, patterns of thought and imaginative models in different ages and cultures. This topic has been discussed in some detail by Dr. Mary Hesse in her book **Science and the Human Imagination** (1954) in which she concludes that science has never been entirely divorced in the past from its cultural and religious environment, and scientific research has room for the creative imagination and for people sensitive to appropriate cultural analogies.

It was no accident that modern science began in the same period as modern Humanism, modern technology and the Age of Reason. The question now is: what are we to think of scientific knowledge now that its partial, limited character is exposed, and its relation to human imagination and culture is beginning to be understood?

4
The Language of Religion

In the case of religion, and particularly Christianity, a parallel situation with regard to language is found. As in every area of science, there is in religious thought an interaction of experience and interpretation. The Gospels in the New Testament present the story of Jesus Christ from different points of view, reflecting the convictions and concerns of the writers. The episodes of the Gospels are interpreted as they are recorded—even more than is done in a newspaper. In the history of Christian thought, there are data of revelation, scripture and experience, and also patterns of interpretation of this data which change with each age or culture. Some patterns of religious interpretation are called mythological—that is they are closely bound to a particular world-view. This may involve a flat universe, with God and his angels walking about on it and encountering men, as in the case of the early stories in the Old Testament. It may be a more developed view with a three-decker universe (heaven, earth and hades) of the kind assumed in other parts of the Old Testament and by many of the Medievals. Or it may be a world-view based on Ptolemaic or Newtonian cosmology found in many hymns still in use. All mythological patterns, according to Professor van Peursen, assert the involvement of man in the situation, and present religious truth in the form of stories which state that spiritual reality is an immediate fact.

Another class of interpretations, again according to van Peursen, attempt to objectify religious reality in terms of special entities, properties or qualities. In the 'classical' period of Christian theology, when the Early Church had to come to terms with the thought of the Greek world, the categories of 'substance', 'essence', 'nature' and so on were applied to God, the person of Christ and the Spirit. In a magnificent synthesis, the Fathers of the Church expressed their understanding of the data of faith in such concepts, and the formulae relating them were hammered out in years of controversy and councils.

More recently, the pattern of interpretation has picked up the relational features of New Testament thought on one hand, or become linked with evolutionary ideas on the other. This has led to theologies of personal encounter associated with the names of Karl Barth and Emil Brunner, to existentialist theology such as that of Rudolph Bultmann, or to process theology like that of de Chardin, Charles Hartshorne and Lionel Thornton. Thus the patterns of religious understanding have changed with changing cultures just as patterns of scientific understanding have done.

As with scientific knowledge, religious thought issues in practical consequences; it also has a community sharing similar attitudes in which understanding is passed on and tested. The literature of religion parallels the literature of science in some degree also.

There are, however, key differences. Genuine forms of religion assert that the reality which meets men most directly and most personally, is that of God. He is, as St. Augustine remarked, closer to us than our senses or our thoughts; he meets us as responsible individuals in the **wholeness** of our lives. Religious experience is the awareness of this reality. It is reflected in a response of awe and wonder of the kind displayed by Isaiah in the Temple at Jerusalem or Peter by the sea of Galilee, or even Charles Darwin contemplating the flowers in his garden in Kent. Further, religious experience issues in commitment to certain attitudes towards people and the world, as well as in specific actions. Such attitudes and actions are guided by a distinctive and overwhelming vision of God which either comes to men immediately (as in mystical experience) or through their experience in the world.

Some of the corporate aspects of Christianity, and its relation to the particular history of the people of Israel, Jesus of Nazareth and the Church of the apostles, are not paralleled in science. The unique and controlling nature of these events, and their interpretation in the community of the church have no counterpart in scientific experience; neither have the activities of prayer, worship or the sharing in the sacraments.

Nevertheless, the task of understanding in religion has led to the use of language with symbols, analogies and models. Symbols (or images) such as the King, Light, Temple, Lamb, the Cross, Fire and Cloud are commonplace in religious language. They lack the precision of mathematical or other scientific symbols, but on the other hand they gain in power by association with historical events and general human experience. Thus they behave like poetic symbols in certain respects; evocative, imaginative, touching our deepest feelings. Thus the phrase 'the light of the world' has an impact and evokes universal response because everyone knows about light and its function of making things clear and safe. Everyone often experiences the world as a mysterious, hostile environment in which it is easy to stumble and get hurt. We also know the world as the sum total of people in need of leadership and unity. The image pictured by this symbol is one which reaches to the depths of ourselves and our situation. When it is linked to the figure of Jesus, the meaning of the Incarnation is communicated in a powerful way, and far more effectively than by a longwinded, systematic exposition of the same thing.

To speak in terms of height or depth about God's otherness—or transcendence—is to use an analogy to express something which would otherwise require a complicated and rarefied language, accessible only to professional theologians. Jesus' use of the word 'Father' when speaking of God is another vital analogy. It does not of course imply physical parenthood of the kind attributed to Zeus, the 'father' of the Greeks and their gods; in this respect the analogy breaks down. But it does reveal to everyone with the experience of fatherly care something of the mysterious relationship made available to the human race by God through Jesus. A parable may be regarded as an extended analogy, once again having wide appeal and obvious limitations.

Models, such as the Body for the Church, Jerusalem for heaven, Adam for Christ or Baptism for Christian life, abound in the New Testament. In particular, the use of human terms to speak of God involves the model of personal life and personal relationships. Prayers are full of this kind of model as a 'way in' to the Christian understanding of the relationship between God and man. This is justified not only by Jesus' use of such language, and by the fact that the personal level is the highest level we humans know, but by Christians' belief that Jesus, in a quite unique way, modelled God for us. As St. Paul put it: 'In him the fulness of the Godhead dwells bodily.' For this reason he is given unique titles; Son of God, Lord, and it is said that: 'God was in Christ reconciling the world to himself.' This personal model, as we have already seen, is reflected in the similarity in language-behaviour between personal and theological discourse.

As in earlier periods of science, the language of religion has often been taken literally. Symbols have replaced the realities to which they referred; analogies have been overpressed; models have been interpreted as final and complete pictures of spiritual relationships. Although the ancient Israelites were well aware of the dangers of graven images, their pious successors have not always taken the warning to heart where human thought and understanding were concerned. Here the Reformed churches have been as much at fault as the Roman Catholic and Orthodox Communions, and the Jew as wayward as the Mohammedan. The religious wars and controversies of Christendom may often be traced to a failure to appreciate the nature of religious language. This, coupled with worldly notions of authority and power, often led to demoniacal acts.

Many of the difficulties of non-Christians about Christianity result from the confusions and absurdities resulting from literal understanding of religious language.

In the past there have been phases of biblical fundamentalism —literal belief in the bible, dogmatic fundamentalism, and even fundamentalism about certain rather obvious images, symbols and models like the cross or the body. The explanations of religion, whether contained in the pages of the bible, the creeds or the devotional vocabulary of the Church, have been taken as final and exclusive. This was particularly true in the latter half of the nineteenth century in the heyday of literal belief in the scientific language of classical physics. Happily, there is a growing recognition today of the limits of religious language, in the bible and outside it. The literal belief in the scriptures is not upheld by the mainstream churches, though it is still widespread among uninformed church members. In the light of the bigotry of the past, few things matter more than that responsible Christians recognize the dangers of literalism, and seek to understand better the relationship between religious language and spiritual reality.

Yet here, as in modern science, there are several different views. The positions of Positivist, Instrumentalist, Idealist and Realist are all found among contemporary theologians as well as among scientists.

The **Positivist** notion is that immediate personal experience is the stuff of religion. This is expressed by various terms of a personal type in the bible, doctrine, liturgy or existentialist philosophy. Theology then becomes a way of making sense of these experiences, but it has no power in itself, being no more than an intellectual interpretation. Revelational and existentialist theologies fall into this category.

The **Instrumentalist** view is that theology provides a map for the religious life. For example, the meaning of righteousness given in the bible, together with the insights of the Church provide the clues for good living. These provide the 'co-ordinates' of a theology which are then applied to practical situations. The symbols and analogies of religion are useful general notions, related to the data of revelation and our experience. Even though they are not precise descriptions of the experiences themselves, they enable us to talk about the mysteries of religion, and provide us with important terms which can shape our thoughts, attitudes and living. Above all on this view, religious language indicates lines of action: theology tells us how to live. Much modern biblical theology is of this kind.

The **Idealist** view is that Christianity is an interpretation of the best life of man before God. Religious language expresses this interpretation, and is essentially an achievement of the human mind, with reference to the life and work of Jesus. The religious mind thus imposes a pattern of meaning upon the vast range of religious experience and example available, even on the many views of Jesus himself found in the changing periods of western culture. Theology is thus a process of abstraction and rationalization with reference to Christ and the insights of the Apostolic Church: its language is the appropriate convention for such discussion. This view has historical links with Medieval nominalism, and is found in liberal and modernist circles.

The **Realist** view, on the other hand, is akin to Medieval phenomenalism. It insists that religious language offers a description of reality, however inadequately. The idea of God as King, for example, is an effective symbol of God's control of the universe. The notion of sacrifice—as it is spelt out by Jesus—provides a descriptive model of God's way of dealing with evil, and of men's proper relationship to him. The theories of the Atonement developed in the history of the Church are only theories, but they are real in so far as they give expression to an aspect of reality which can in the nature of the case only be understood partially in any age. Modern catholic theology is of this type.

In view of these remarkable parallels with the philosophy of science, it is not surprising to notice that theology today allows for an element of **indeterminacy**—and not only in its 'fine structure' as in modern physics. Because of the nature of religious reality, and the limits of human thought and language, there exists of necessity a **complementarity of religious models and analogies for any satisfactory understanding.** This is confirmed by an examination of the many parables, theories, concepts and models in the Old and New Testaments, and by the extensive use of poetic and personal language throughout the bible.

What religious language lacks—and scientific language necessarily possesses—is meaning given by experiment under controlled conditions. The only test available to men of faith is a complete commitment to God, which is different in kind from scientific assumption and experimentation. On the other hand, Christianity, like Judaism and Islam, lays claim to an historical and unrepeatable foundation. This provides the basic clues and framework of the theological language-map. These differences pin-point the differing kinds of knowledge in science and religion, and the different place of faith and doubt in scientific and religious attitudes.

The languages of science and religion therefore reveal the similarities and dissimilarities of these two kinds of human experience. They show the reasons for conflict when the languages, terms or attitudes are confused, and the possibilities of reconciliation when the nature of each is understood. This discussion points to the need for mutual respect, and the acceptance of the limitations of both science and religion. But inevitably, it raises further questions: what is the relation between faith and doubt in religion and science? What precisely do we mean by scientific knowledge or religious knowledge? What kind of evidence is convincing for religious understanding, and how can it be shown to be reliable? How can the realms of science and religion be related? An answer to these questions requires a study of worship, wonder, commitment and faith, in science and in religion.

5

Topics, Questions and Sources

1

Lord Ritchie Calder spoke of the growing problem of communication among specialists working in different scientific areas (page 109). Collect examples of scientific 'jargon' to illustrate this. What do you know about the new language systems for computers?

2
Why do you think people look for a single pattern of explanation for all kinds of experience?

3
Describe Wittgenstein's picture theory of language.

4
What is a parable? How does it operate? Try to write a modern one.

For Discussion:

5
'Is it possible to verify a religious belief?'

6
'Scientific objectivity is only the agreement of other scientists.'

7
'The poet and the craftsman, by the nature of their work, can know more of God than the philosopher or theologian'.

Sources:

Issues in Science and Religion I. G. Barbour *S.C.M.* 1966 *Pt. II.*
Science and Religion H. K. Schilling *Allen & Unwin* 1963.
Science and Religion: Conflict and Synthesis I. T. Ramsey *S.P.C.K.*
The Philosophy of Science S. Toulmin *Harper P.B.* 1960.
The Philosophy of Religion John Hick *Prentice-Hall* 1965.
Ludwig Wittgenstein W. D. Hudson *Lutterworth Press* 1968.
Models and Mystery I. T. Ramsey *O.U.P.* 1964.
Language, Logic & God Frederick Ferré *Eyre & Spottiswood* 1962.

D
The Impact of Science on Theology

No discussion of the modern debate between science and religion would be complete without some reference, however brief, to the influence of scientific achievements and attitudes on religious understanding. This influence has been exerted in several different ways, some directly and others in a more oblique fashion.

1
The Problem of Miracles

We have noted already the way in which the stress on human reason was a major element in the rise of modern science, and the world-view of Newton's time. Accepting both the premises and the conclusions of Newton's scheme, large numbers of churchmen believed that God should be understood as the Designer of the world-machine, and its originator. The miraculous elements in the biblical record were explained away by rationalist commentators. Christianity was presented as a religion of reason, and was, in the words of John Toland's book, **Christianity not mysterious** (1696). Evidence for God was found not in history

R.S.—I

but in the design of nature and the cosmos as it was then understood. Increasingly, God's continuing role and activity were ignored. In place of the living God of the bible, men were left with the remote God outside the machine. These Deistic views soon ran out into Agnosticism and Scepticism, and were attacked both by Churchmen like Bishops Butler and Berkeley, and by sceptical philosophers such as David Hume. The problems of miracle and prophecy have, however, always remained key issues in the interaction of science and religion.

Traditionally, a miracle (latin: *miraculum*) has been thought of as a break in natural law. Religious thinkers argued for the possibility of such discontinuities on the grounds that the scriptures assert that they occurred, faith has been built (in part at least) upon them, and that if the laws of nature are God's laws, there is no reason why he should not suspend them for his own purposes.

Hume cast doubt on this view by several counter-arguments. First, by saying that wise men base their judgement of such stories on their own experience and on probability. Against the testimony of possibly credulous people of a long time ago must be set our universal experience of the law of nature, and reasonable people will prefer to conclude that some mistake was made rather than accept a quite extraordinary event like that of a man rising from the dead. Secondly, the testimony of miracles is not given by men of such good sense or education as to encourage us to believe in it without hesitation. Thirdly, there is a natural attraction to and sensationalism about miraculous events which should make us wary of such reports. Fourthly, different religions appeal to the same miracles, and thus discredit each other and their miraculous foundations!

The attack is impressive, but at certain points it fails. However unlikely a miracle may be, it cannot be ruled out as a possibility, even though we may never have ourselves had experience of one. A law of nature is no more than the sum total of all previous experiences. Predictions can be made on the basis of such laws only on the assumption that nature is uniform. We can therefore say no more than miracles are improbable or uncommon. They cannot automatically be ruled out as impossible: such a conclusion would be quite unscientific and illogical. Moreover, if a miracle is an historical event, it is in the nature of such events to be unique and unrepeatable. Thus a miracle may prove an exception to a law of nature without destroying the general usefulness of that law. Only a repeatable experiment carried out under controlled conditions could do that; and the last thing a miracle seems to be is a repeatable experiment!

As to Hume's points about the need for good evidence before believing in a miracle, he was clearly on good ground. Historical investigation of the most rigorous kind is necessary to establish any claim that an event was in fact quite extraordinary. The credibility of the witnesses must be carefully examined, and there are considerable difficulties about doing this on ancient documents coming to us from a very different, non-scientific world; the difficulties are acute enough **in recent** documents, such as those concerning the assassination of President John Kennedy in 1963, for example. Nevertheless, the work of biblical scholars over the past hundred and fifty years has provided a very rigorous examination of the testimony and the witnesses to events such as the resurrection of Jesus. The great majority undoubtedly subscribe to the view that a quite shattering event did occur after his death, which was preceded by a series of very unusual and significant happenings. And they would agree with David Hume himself, who ended his famous **Essay on Miracles** with the words:

> *The Christian religion not only was at first attended with miracles, but even at this day cannot be believed by any reasonable person without one. Mere reason is insufficient to convince us of its veracity: and whoever is moved by faith to assent to it is conscious of a continued miracle in his own person, which subverts all the principles of his understanding, and gives him a determination to believe what is most contrary to custom and experience.*

Hume here recognizes that a miracle can have another meaning than an external, extraordinary event. He also hints that it belongs to the realm of faith or religious insight—which is the same as prophetic understanding—rather than merely to the material world. This accords with the more recent view of a miracle, which is understood as an event disclosing God to those with faith. On this definition, 'miraculous' events in the past for which explanation may now be given (such as the exodus or the feeding of the multitude by Jesus) remain miracles in the prophetic sense, as events which disclose God's nature and will to those with eyes to see it. Thus even in the unlikely event of the bones of Jesus being found somewhere in Palestine—a possibility suggested by Professor Ronald Gregor Smith in his book **Secular Christianity**—the significance of the resurrection as a spiritual event, a miracle in this new sense, would be largely untouched.

2
Studying the Bible

Scientific achievement has affected Christian understanding also in the realm of biblical studies. Archaeological investigations have thrown considerable light on the background of the ancient Near East, the history of the people of Israel, and the various religious, political and cultural influences which helped to shape the world from which the bible came. The investigation of semitic languages and discovery of other ancient writings such as the Dead Sea scrolls have enabled scholars to make considerable advances in understanding the biblical texts.

Between 1929 and 1933, C. F. A. Schaeffer discovered a collection of ancient Canaanite tablets at Ugarit (Ras Shamra) on the coast of northern Syria. The find proved to be a sensational one, containing much religious literature of the Canaanite (that is, Phoenecian) culture. Work by Charles Virolleaud and H. L.

Ginsberg on these texts revealed many parallels with Hebrew language and literature, throwing much light on Hebrew ways of thought and writing. Parallels in the epic stories, names of heroes and mythology also abound, indicating the common inheritance of Hebrew and Canaanite literature—but at the same time, the distinctive insights and convictions of the Hebrews emerge with great clarity.

Textual and literary criticism has made great strides in the scientific era, employing scientific methods of obtaining evidence, formulating hypotheses and testing them. An example concerns variants in certain texts of St. Mark's Gospel, chapter 1.41—an episode about a leper who came to Jesus. The majority of ancient texts when translated read 'he (Jesus), being moved with compassion.' A smaller number read 'he, being moved with anger.' This reading is preferred because the evidence is that more difficult readings are more likely to be original, and the trend of interpretation is invariably to reconcile hard sayings with the author's general convictions—in this case of Jesus as a man of compassion. Here a scientific process of dealing with evidence, forming a theory and testing it in translation is in use.

Newer forms of criticism, like form criticism—in which account is taken of the form of sections of biblical literature in relation to the social setting for which it was prepared, and the convictions of the writer—have opened up new ways of understanding and interpreting the material. The variety, interest, complexity and significance of the bible have emerged from this critical investigation in a most exciting way, having quite new relevance to modern ways of thought and life.

3
The Challenge of Sociology

Sociology has also made its contribution to Christian thought. Not only has the social reference of the biblical literature, with its social models and symbols like the city, the nation, and kingship come to light, but the new study of the sociology of religion is indicating the role of religion in societies of different kinds, and revealing hitherto unrecognized factors affecting Christian belief and practice.

Emile Durkheim, the great French sociologist, concluded from his study of certain primitive societies that the gods whom men worship are representations of the form of society to which they belong rather than supernatural realities. The group exercises the power of a god in relation to individual members and this power creates the idea of god who is a symbol of the society.

There is no doubt that society has a reality over and above the sum total of its individual members. It can claim loyalty, devotion and self-sacrifice, especially in times of war. Each society does seem to have an effect on certain features of the idea of God held by its members. Do not most Englishmen think of him as a super clergyman with a white face? But Durkheim's theory, while it may have some application in primitive communities, certainly does not account for all the facts of religion as they have come down to us from the biblical revelation or Christian history. Christianity goes beyond the limits of any one society or grouping. Jesus is believed to be the leader of a new humanity, not merely a new Judaism. Again, Durkheim's theory provides no place for those who stand against society in the name of God like the Old Testament prophets. Nor does it allow for the phenomenon of conscience in, for example, those who out of religious conviction opposed the Nazi regime in Germany during the last war. As in previous naturalistic explanations, this kind of theory may give a useful indication of **origins** of religious dispositions, but cannot evaluate the **validity** of religious belief or explain it away in sociological terms. This is the reductionist fallacy again.

Today, sociologists are increasingly concerned with the social phenomena of religion as 'sociological facts'. Recent studies have shown the surprising extent of occasional church-going in England, and the key place in the suburban community of that professional amateur, the parish priest or minister. Other studies have revealed the long estrangement of the churches from the industrial workers in large towns, and pointed to the very different style, culture and language of the churches from that of large sections of the population. It seems a curious fact that the pattern of church-going—at 8 a.m., 11 a.m. and 6.30 p.m.—found appropriate for the eighteenth-century village community should still be thought suitable for urban churches in the age of the motor car and television. And it appears a little odd that the Latin-based English of three hundred years ago should be thought the right language for a mid-twentieth century Englishman to use when addressing the Almighty! Studies of these and other religious phenomena by sociologists are gradually forcing Christians to reconsider the meaning of their faith, and encouraging a number of reforms in religious practice.

4
Speaking of God

It is in the area of basic theology that the backwash of scientific understanding is being most disturbingly felt.

Following his experience in the 1914–18 war as a German Army Chaplain, Professor Rudolph Bultmann concluded that the Christian Gospel had to be extracted from the mythological framework of the first-century world if it was to be communicated to scientific man. At the same time, Bultmann proposed that the categories fundamental to this reinterpretation of the New Testament message were those of the existentialist philosophy of Professor Martin Heidegger. Heidegger was a colleague of Bultmann's at the University of Marburg from 1923 to 1928.

Existentialism, of which Heidegger's is one form, stresses the personal 'subject to subject' relationship of what it calls 'authentic human existence' in contrast to the subject-object relationship of scientific thought. It makes an analysis of existence in human terms, rather than in terms of things, as traditional philosophy has done. It is to be seen in the writings of Jean Paul Sartre, Albert Camus and Iris Murdoch, as well as in much modern art and drama.

Bultmann asserted that the New Testament understanding of faith was basically that expressed by his extension of Heidegger's philosophy, and proceeded to interpret the Christian message in the light of it. This is his programme of demythologizing, most clearly outlined in his book **Jesus Christ and Mythology**. The programme entailed an elimination of miracles as historical events—understanding them instead as mythological ways of expressing religious insights appropriate to the early Church and its world-view. The resurrection of Jesus is treated as an affirmation of faith on the part of Christians that Jesus is the 'authentic Word of God to man' rather than a scientific or historical statement of what happened. Faith is thus emphasized as a subjective, inner, self-justifying attitude rather than a response to verifiable, historical facts.

Bultmann was prepared to preserve the essential language and content of the New Testament preaching—the *kerygma*—as it has been delivered from apostolic times, recasting the surrounding teaching of the evangelists or St. Paul. Some of his followers have carried his programme still further, attempting to reinterpret the heart of the message itself in existentialist terms, and arousing bitter opposition from more orthodox Christians in the Church. This whole situation has come about as a direct result of the impact of scientific thought, especially of the logical positivist phase in the 1920's. In the last 50 years, Christian theologians have stressed the personal, non-verifiable aspects of faith at the expense of a reliance upon historical evidence or general experience in the world—including the scientific experience. There are signs now, however, that this trend is being reversed in the light of recent philosophical investigations.

In **Honest to God**, (1964) Bishop John Robinson collected brief summaries of theologies of several Christian thinkers influenced by the existentialist movement: Tillich, Bonhoeffer, and Bultmann. He used these to suggest that the 'traditional' mythology of popular Christianity with God above or 'up there' needed reappraisal, suggesting that the imagery of depth, human love and self-awareness should replace it. He was attacking the understanding of God generated by the rationalistic world-view of Newtonian physics, still widely preached by the Church. His premise for doing so was the changing world-view of modern science. The resulting reaction in the Church was a salutary experience! Yet by means of mass media of communication in publishing and television, the attention of the world at large was caught by a theoretical issue for the first time for many years. 'Scientific man' showed that he had not completely lost interest in the questions of theology.

Even more searching and radical have been the theological debates since 1965 about the meaning of the word 'God' and the relevance of the concept to Christianity. This has led to various extreme theologies of secular Christianity such as those of Dr. van Buren and the 'Death of God' school in the United States of America. Further light has been thrown upon the meaning of the key words of theology by biblical and patristic studies (i.e. the writings of the early fathers of the Church up to the end of the fifth century A.D.). It has been pointed out that the categories in which the western Church has thought of God in the past centuries have shifted from those of the early Church set in the Greek world, and even more from the very different world of the biblical writers, which are certainly not those of the modern scientific era.

In the bible, the words for God are defined by the meaning of the key events which created the people of Israel. The understanding of the word is given enigmatically in the great passage in Exodus 3 when God's name is disclosed to Moses as JHWH—'I will cause to be'. His mysterious nature, will and purpose are spelt out in the covenant relationship sealed in sacrifice on Mount Sinai under Moses. Israel's understanding of it is further developed in relation to the covenant and the continuing history of the people of Israel by the prophets, and expressed in the people's worship, social life, key figures and institutions. From what happened in Israel's **history**, God came to be understood as personal, righteous, holy, good, merciful; the mysterious Lord of history and creation; the everpresent Spirit of love, justice and creative power. And all this was inferred from historical events interpreted in the light of the dynamic relationship of Israel with her covenant-God.

Today it is suggested that scientific understanding goes hand in hand with an awareness of mystery in the world of matter, energy and space. It is also emphasized that science is built upon functional relationships between the observer and what is observed, and between elements in the picture presented by scientific language. Thus we determine speed in relation to time and distance, or distance in relation to speed and time. In other words, the sciences show the relations **between** things rather than all there is to know about things **in themselves**. A key example is Einstein's theory of relativity which abandons the idea of absolute space and time. Because of this it is argued that the key concepts of religion need to be recast into relational or functional categories. This raises some very profound questions. Is the ultimate reality to be found in some sort of relationship between what is within and what is without us?

The idea of a God who 'acts', 'speaks', 'loves' and is known in relation to people or the creation is thus strongly commended, over against the traditional terms of 'being', 'existence', 'essence', or divine 'substance'. Similarly, attention is directed away from the two-nature model of Christ—and the unresolved problem of how God and Man can be united in one person—towards the function of Jesus as life-bringer, revealer or exemplar of love, through whose humanity God's relation to us is perfectly expressed. Renewed interest is being shown in the theology of the Holy Spirit—for long relegated to an inferior place in the western Church. As we have seen, the distinction between scientific and theological language gives new prominence to the idea of the Spirit. God's obviously indefinable, spiritual nature—

yet felt activity—chimes in well with what we know of the nature of religious language, and serves to make the necessary contrast with the realities of the material world investigated by science.

In ways like these, the scientific revolution is continuing to influence religious thought, forcing Christians and men of other faiths to refine their understanding and make it communicable in modern society. The Church is left with fascinating and exciting new problems: how shall we speak of God in a scientific age? Is it possible to avoid speaking of God in terms of being or existence? What language will communicate best the realities of faith to scientific man?

5

Topics, Questions and Sources

Apart from those questions mentioned at the end of the last section, the following can be considered:

1
Debate the motion: 'In the opinion of this house, miracles never happen'.

2
Do you think it would make any difference to Christian faith if the bones of Jesus were found in Palestine?

3
Prepare an exhibition indicating some of the ways in which archaeology is influencing the understanding of the bible.

4
What are the following: textual criticism, source criticism, form criticism as applied to the bible?

5
Explain why knowledge of the origin of a belief can give no indication of the truth or falsehood of that belief.
6
As a group project, find out as much as you can about the meaning of the word 'God', and the way that meaning has emerged, from primitive times through to the New Testament period.

7
To the Africans Christ is a 'black' man, to the Asians, he is a 'yellow' man, to the Europeans, a 'white' man. Is this inevitable, and is it a good thing to think in this way?

Sources:

Philosophers and Religious Truth R. N. Smart *S.C.M.* 1964.
The Bible in the Age of Science A. Richardson *S.C.M.* 1961.
The Archaeology of Palestine W. F. Albright *Penguin* 1963.
The Philosophy of Religion (Sociological Reductionism—Ch. 3) John Hick *Prentice-Hall* 1965.
Jesus Christ and Mythology R. Bultmann *S.C.M.* 1960.
Rudolph Bultmann Ian Henderson *Lutterworth* 1965.
Paul Tillich J. Heywood Thomas *Lutterworth* 1965.
Dietrich Bonhoeffer E. H. Robertson *Lutterworth* 1965.
Martin Buber R. Gregor Smith *Lutterworth* 1966.
The Secular City (Ch. 11) Harvey Cox *S.C.M.* 1965.
Christian Theology and Metaphysics P. R. Baelz *Epworth* 1968.

Index

date
ease